STUDENT'S STUDY GUI
SOLUTIONS MANUAL
VOLUME 3: CHAPTERS 37–44

SEARS & ZEMANSKY'S

UNIVERSITY
PHYSICS

15TH EDITION

WAYNE ANDERSON
A. LEWIS FORD

With contributions by Joshua Ridley

330 Hudson Street, NY NY 10013

Director Physical Science Portfolio: Jeanne Zalesky
Physics and Astronomy Portfolio Analyst: Ian Desrosiers
Senior Content Producer: Martha Steele
Managing Producer: Kristen Flathman
Courseware Editorial Assistant: Sabrina Marshall
Director, Mastering Physics Content Development:
Amir Said

Associate Producer, Science/ECS: Kristen Sanchez
Senior Content Developer (Physics/Astronomy):
David Hoogewerff
Rich Media Content Producer: Keri Rand
Full-Service Vendor: SPi Global
Compositor: SPi Global
Manufacturing Buyer: Stacey Weinberger

Cover Photo Credit: John Woodworth/Alamy Stock Photo

Printed in the United States of America.
1 2019

ISBN 10:   0-135-59202-X
ISBN 13: 978-0-135-59202-1

CONTENTS

Part VI   Modern Physics

This student resource to accompany Volume 3 (Chapters 37–44) of the fifteenth edition of *University Physics* by Roger Freedman and Hugh Young has been updated to combine two separate resources into one text: the *Student's Study Guide and Solutions Manual*.

The *Study Guide* summarizes the essential information in each chapter and provides additional problems for the student to solve, reinforcing the text's emphasis on problem-solving strategies and student misconceptions. It now also includes the new Key Concept Statements from each Worked Example in the text and tags these concepts to the Conceptual Questions and Problems presented in the *Student's Study Guide*. Finally, it includes the Key Example Variation Problems from the text, new to the Fifteenth Edition.

The *Solutions Manual* contains detailed solutions for all of the new Key Example Variation Problems and approximately two-thirds of the odd-numbered Exercises and Problems in Chapters 37–44. The Exercise and Problems included were not selected at random but rather were carefully chosen to include at least one representative example of each problem type. The solutions are intended to be used as models to following in working physics problems, and are worked out in the manner and style in which you should carry out your own problem solutions. The remaining Exercise and Problems, for which solutions are not given here, constitute an ample set of problems for you to tackle on your own.

The *Student's Study Guide and Solutions Manual* Volume 1 (Chapters 1–20) and Volume 2 (Chapters 21–37) are also available from your college bookstore.

Wayne Anderson
Lewis Ford
Sacramento, CA

37

STUDY GUIDE FOR RELATIVITY

Summary

In this chapter, we study the special theory of relativity introduced by Einstein in 1905. The theory is based on two postulates: The laws of physics are the same in every inertial reference frame, and the speed of light in vacuum is the same in all reference frames. These simple postulates have far-reaching implications. We'll explore how two observers moving relative to each other might not measure the same time or length, how two events might not be simultaneous in all reference frames, and how we must modify the principles of momentum and energy to suit special relativity. You'll find that your intuition is often unreliable in these types of situations. We'll learn to develop new tools to analyze relativity problems.

Objectives

After studying this chapter, you'll understand

- The two postulates of Einstein's special theory of relativity.

- How to identify inertial frames of reference.

- When to apply the special theory of relativity.

- The definition of *proper time* and *proper length*.

- How to apply time dilation and length contraction to various problems.

- How to use Lorentz transformations to find quantities in various reference frames.

- The concepts of, and how to apply, relativistic momentum and relativistic energy.

- The concept of rest energy.

From Chapter 37 of Student's Study Guide to accompany *University Physics with Modern Physics, Volume 3,* Fifteenth Edition.
Hugh D. Young and Roger A. Freedman. Copyright © 2020 by Pearson Education, Inc. All rights reserved.

Concepts and Equations

Term	Description
Relativity and Simultaneity	All fundamental laws of physics have the same form in inertial reference frames. The speed of light in vacuum is the same in all inertial frames of reference and is independent of the motion of the source. Simultaneity is not absolute: Two events occurring simultaneously in one frame might not appear simultaneous in a second frame.
Time Dilation	The proper time Δt_0 is the time interval between two events that occur at the same spatial point in a frame of reference. If this frame moves with a constant velocity u relative to a second frame, the time interval Δt between events observed in the second frame is longer: $$\Delta t = \frac{\Delta t_0}{\sqrt{1 - u^2/c^2}} = \gamma \Delta t_0,$$ $$\gamma = \frac{1}{\sqrt{1 - u^2/c^2}}.$$ This effect is known as time dilation.
Length Contraction	The proper length l_0 is the distance between two points at rest in a frame of reference. If this frame moves with a constant velocity u relative to a second frame, the distance l measured parallel to the frame's velocity in the second frame is shorter: $$l = l_0 \sqrt{1 - u^2/c^2} = \frac{l_0}{\gamma}.$$ This effect is known as length contraction.
Lorentz Transformations	Lorentz transformations relate the coordinates and time of an event in an inertial coordinate system S to the coordinates and time of a second inertial frame S' moving with constant velocity u relative to the first frame. The Lorentz transformations are $$x' = \frac{x - ut}{\sqrt{1 - u^2/c^2}} = \gamma(x - ut), \qquad y' = y, \qquad z' = z,$$ $$t' = \frac{t - ux/c^2}{\sqrt{1 - u^2/c^2}} = \gamma(t - ux/c^2).$$ For one-dimensional motion, the velocities in the two systems are related by $$v_x' = \frac{v_x - u}{1 - uv_x/c^2},$$ $$v_x = \frac{v_x' + u}{1 + uv_x'/c^2}.$$
Doppler Effect for Electromagnetic Waves	The Doppler effect is a shift in the frequency of light from a source that is moving relative to an observer. For a source moving toward an observer at speed u, the frequency shift is given by $$f = \sqrt{\frac{c + u}{c - u}} f_0.$$

Relativistic Momentum and Energy	The relativistic momentum and energy are respectively given by
	$$\vec{p} = \frac{m\vec{v}}{\sqrt{1 - v^2/c^2}} = \gamma m \vec{v}, \qquad K = \frac{mc^2}{\sqrt{1 - v^2/c^2}} - mc^2 = (\gamma - 1)mc^2.$$ The total energy of a particle is the sum of its kinetic energy and its rest mass energy $E_{\text{rest}} = mc^2$, given by $$E = K + mc^2 = \frac{mc^2}{\sqrt{1 - v^2/c^2}} = \gamma mc^2.$$ The total energy can also be expressed as a function of momentum and rest mass: $$E^2 = (mc^2)^2 + (pc)^2.$$

Key Concept 1: The time interval between two events depends on the frame of reference of the observer. The time interval is shortest in an inertial frame S where the two events occur at the same point; this is the proper time Δt_0. In any other inertial frame that is moving with respect to S, the time interval is longer (it is *dilated*) [Eq. (37.6)].

Key Concept 2: If two events take place at the same point in an inertial frame S, an observer in a frame S' moving relative to S observes a time interval between these events that is greater than in S by the Lorentz factor γ [Eqs. (37.7) and (37.8)]. This factor is substantially different from 1 only if the relative speed of S' and S is a substantial fraction of the speed of light.

Key Concept 3: No matter which inertial frame of reference you are in, you observe dilated time intervals for timekeepers (clocks) of any kind that move relative to you. If S and S' are two inertial frames in relative motion, an observer in S measures clocks in S' to run slow, and an observer in S' measures clocks in S to run slow.

Key Concept 4: The dimensions of an object depend on the frame of reference of the observer. If the object is moving at constant velocity relative to your inertial frame, its length is contracted *along* the direction of relative motion [Eq. (37.16)]. There is no change in length *perpendicular* to the direction of relative motion.

Key Concept 5: No matter which inertial frame of reference you are in, you observe objects that move relative to you to be contracted along the direction of relative motion. If S and S' are two inertial frames in relative motion, an observer in S measures objects in S' to have contracted lengths, and an observer in S' measures objects in S to have contracted lengths.

Key Concept 6: Given the position and time of an event as measured in one inertial frame of reference, you can use the Lorentz coordinate transformation to find the position and time of that same event as measured in a second inertial frame moving relative to the first one.

Key Concept 7: Given the velocity of an object as measured in one inertial frame of reference, you can use the Lorentz velocity transformation to find the velocity of that same object as measured in a second inertial frame moving relative to the first one. If an object is moving slower than c as measured in one inertial frame, its speed is less than c as measured in any other inertial frame.

Key Concept 8: The frequency of electromagnetic waves as measured by an observer moving relative to their source is different from the frequency measured in the rest frame of the source. The measured frequency is greater than the source frequency if the observer and source are approaching, and less than the source frequency if they are receding [Eq. (37.25)].

Key Concept 9: The expression for momentum [Eq. (37.31)] and the relationship between net force and acceleration [Newton's second law, Eqs. (37.32) and (37.33)] are both modified in the special theory of relativity. As the speed of a material particle approaches c, its momentum approaches infinity and the acceleration caused by a net force approaches zero.

From Chapter 37 of Student's Study Guide to accompany *University Physics with Modern Physics, Volume 3,* Fifteenth Edition.
Hugh D. Young and Roger A. Freedman. Copyright © 2020 by Pearson Education, Inc. All rights reserved.

Key Concept 10: The expression for kinetic energy [Eq. (37.36)] is modified in the special theory of relativity. As the speed of a material particle approaches c, its kinetic energy approaches infinity. The total energy of a particle of rest mass m is the sum of its kinetic energy and its rest energy mc^2.

Key Concept 11: In a relativistic collision, momentum and total energy are both conserved. Particles can be created or destroyed in such collisions by converting kinetic energy to rest energy, or vice versa.

Conceptual Questions

1: When does relativity become important?

A train is speeding past a platform. At what speed must the train be traveling for the proper time measured on the train to differ from the time measured on the platform by 0.1%?

IDENTIFY, SET UP, AND EXECUTE The time dilation relation will solve this problem. The proper time (Δt_0) on the train will transform to the time measured on the platform (Δt) by

$$\Delta t = \frac{\Delta t_0}{\sqrt{1 - v^2/c^2}}.$$

We want to find the velocity corresponding to a 0.1% difference in time, so we set Δt to $1.001\Delta t_0$:

$$\Delta t = 1.001\Delta t_0 = \frac{\Delta t_0}{\sqrt{1 - v^2/c^2}}.$$

Solving gives

$$v = \sqrt{1 - \left(\frac{1}{1.001}\right)^2}\, c = 0.045c.$$

A 0.1% time difference will occur when the train is moving at $0.045c$, or 1.3×10^7 m/s.

EVALUATE For a very small relativistic change, the train must be moving at over 13 million meters per second, or 48 million kph. It is highly unlikely that we'll see such effects on any train we'll ever travel on.

2: Proper time and length

You are traveling on a spacecraft moving at $0.1c$ past a space station. If you are holding a physics textbook and flashing LED, who measures the proper length of the book and the proper time interval for the flashing LED, you or an observer on the space station?

IDENTIFY, SET UP, AND EXECUTE The proper length is measured in a frame in which the book is at rest. Therefore, you measure the proper length of your physics textbook. The proper time is also measured with respect to a frame at rest. Again, you measure the proper time for the flashing LED.

EVALUATE Identifying the correct frame is often the most challenging step in a relativity problem. Here, the observer on the station would measure a length-contracted physics textbook and a time-dilated flashing LED.

3: Increasing energy

An electron is traveling at a speed of $0.95c$. Can its energy be increased by more than 5%? by more than 25%? by more than 500%?

IDENTIFY, SET UP, AND EXECUTE Relativistic energy for the electron is given by

$$E = \frac{mc^2}{\sqrt{1 - v^2/c^2}}.$$

As v gets closer and closer to c, the denominator gets closer to zero and the energy increases to infinity. For this reason, there is no limit on the maximum energy of the electron.

EVALUATE For the electron to achieve an increase in energy of 500%, it will need to travel at $0.998c$.

Problems

1: Marley and Cassie

Marley, a twin, takes a round-trip journey to a distant star 12 light years from earth at a speed of $0.93c$. Cassie, the second twin, remains on earth. What is the age difference between Marley and Cassie when Marley returns?

IDENTIFY We need to determine the proper time and use time dilation to solve the problem.

SET UP We need to find the time taken for the journey, as observed by both twins. The elapsed time for Cassie, the twin on earth, will be the distance of the journey divided by the speed. The elapsed time for Marley, the traveling twin, will be the elapsed time on earth transformed to the proper time of the traveler.

EXECUTE The elapsed time for the twin on earth (Cassie) is the distance traveled divided by the speed:

$$\Delta t = \frac{c(24 \text{ years})}{0.93c} = 25.8 \text{ years}.$$

Note that the round-trip journey takes 25.8 years. The time in the traveling twin's (Marley's) reference frame is the proper time, since she is the observer at rest. We find this time interval from the time dilation equation:

$$\Delta t = \frac{\Delta t_0}{\sqrt{1 - u^2/c^2}},$$

$$\Delta t_0 = \Delta t \sqrt{1 - u^2/c^2} = (25.8 \text{ years})\sqrt{1 - (0.93c)^2/c^2} = 9.48 \text{ years}$$

Marley ages 9.48 years, while Cassie ages 25.8 years. The difference between Marley's and Cassie's ages will be 16.3 years after Marley returns.

KEY CONCEPT 2 **EVALUATE** This is the classic twin paradox. We see that Marley, the traveling twin, ages much more slowly than Cassie, the stationary twin. Critical to this problem was identifying which frame was associated with the proper time.

CAUTION **Find the Proper Time Carefully:** Determining the proper time is critical in time dilation problems. The proper time is the time measured in a frame that is moving with the clock.

Practice Problem: What is the age difference when Marley's speed is 0.95c? *Answer:* 17.4 years

From Chapter 37 of Student's Study Guide to accompany *University Physics with Modern Physics, Volume 3*, Fifteenth Edition.
Hugh D. Young and Roger A. Freedman. Copyright © 2020 by Pearson Education, Inc. All rights reserved.

Extra Practice: How fast does Marley need to be going so that the age difference is 20 years? *Answer:* 0.984c

2: Moving spacecraft

A spacecraft moves past the earth at a speed of 0.850c. A student on earth measures the length of the moving spacecraft to be 97.0 m. How long does the spacecraft appear to the crew on the ship?

IDENTIFY We'll use the length contraction equation to solve the problem.

SET UP The proper length is the length measured in a frame in which the spacecraft is at rest. The crew will measure the spacecraft when it is at rest, which is the proper length. The student measures the relativistically contracted length.

EXECUTE The length contraction equation is

$$l = l_0\sqrt{1 - u^2/c^2}.$$

The student measures the contracted length and we need to find the proper length. Solving for l_0 yields

$$l_0 = \frac{l}{\sqrt{1 - u^2/c^2}} = \frac{(97.0\text{ m})}{\sqrt{1 - (0.85c)^2/c^2}} = 184\text{ m}.$$

The spacecraft's crew would measure the ship to be 184 m long.

KEY CONCEPT 4 **EVALUATE** This is a classic length contraction problem. The length we found was longer than the measured length because the measured length was the contracted length. You must carefully determine which length is the proper length.

CAUTION **Find the Proper Length Carefully:** Determining the proper length is critical in length contraction problems. The proper length is the length measured in a frame in which the body is at rest.

Practice Problem: What speed is required so that the students measure half the proper length? *Answer:* 0.866c

Extra Practice: What length is measured when the speed is 0.9c? *Answer:* 80.2 m

3: Spaceship relativity

A spaceship moving away from the earth at a speed of 0.60c fires a 5.0-m-long missile in its direction of motion with a speed of 0.20c relative to the spaceship. A crew member on the ship observes that the firing takes 10.0 s. (a) What is the missile's speed relative to earth? (b) What is the length of the missile prior to firing, as observed on earth? (c) What is the time interval of the firing event, as measured on earth?

IDENTIFY AND SET UP We'll use a Lorentz transformation, time dilation, and length contraction to determine the values in the earth's frame.

EXECUTE From the Lorentz transformation, the speed of the missile relative to earth is found to be

$$v = \frac{v' + u}{1 + uv'/c^2} = \frac{(0.20c) + (0.60c)}{1 + (0.60c)(0.20c)/c^2} = 0.71c,$$

where we have set the moving frame (S') to be the frame of the spaceship moving at speed $u = 0.60\ c$ and the speed of the missile to be $v' = 0.20\ c$.

The length of the missile as measured on the earth is found from the length contraction relation. The known length was measured on the spaceship and is therefore the proper length. The length on earth is

$$l = l_0\sqrt{1 - u^2/c^2} = (5.0\text{ m})\sqrt{1 - (0.60c)^2/c^2} = 4.0\text{ m}.$$

The time elapsed for the firing of the missile, as measured on the earth, is found from the time dilation relation. The known firing time was measured on the spaceship and is therefore the proper time. The firing time on earth is

$$\Delta t = \frac{\Delta t_0}{\sqrt{1 - u^2/c^2}} = \frac{(10.0\text{ s})}{\sqrt{1 - (0.60c)^2/c^2}} = 12.5\text{ s}.$$

As measured on earth, the speed of the missile is $0.71c$, the length of the missile is 4.0 m, and the time taken to fire the missile is 12.5 s.

KEY CONCEPT 7 **EVALUATE** This problem combined several aspects of relativity. In contrast to the previous problems, the given time and length were in the proper frame.

Practice Problem: How long is the missile as measured by someone on earth after it is fired? *Answer:* 3.52 m

Extra Practice: How long would the missile be before firing as measured by a member on the spaceship if someone on earth measured it to be 5 m after firing? *Answer:* 6.25 m

4: Accelerating an electron

An electron is accelerated from rest to a velocity of $0.9c$ by a potential difference. Calculate the potential difference. The rest energy of the electron is 0.511 MeV.

IDENTIFY We'll use the relativistic energy to find the solution.

SET UP The change in kinetic energy is the change in the electric potential energy. We'll find the potential needed to create the corresponding change in kinetic energy. Recall that 1 MeV is 1 million electron volts, a unit of energy.

EXECUTE The initial energy of the electron at rest is 0.511 MeV. The energy of the electron at $0.9c$ is

$$E = \left(\frac{m_{electron}c^2}{\sqrt{1 - v^2/c^2}}\right) = \left(\frac{0.511\text{ MeV}}{\sqrt{1 - (0.9)^2}}\right) = 1.172\text{ MeV}.$$

The change in energy is due to the change in electric potential energy. Algebraically, this is

$$e\Delta V = \Delta E$$
$$= E - mc^2$$
$$= 1.172\text{ MeV} - 0.511\text{ MeV} = 0.661\text{ MeV}.$$

A potential difference of 661,000 V is needed to accelerate the electron to $0.9c$.

KEY CONCEPT 10 **EVALUATE** We see how we must combine our knowledge of energy in general with our knowledge of relativistic energy in this problem.

From Chapter 37 of Student's Study Guide to accompany *University Physics with Modern Physics, Volume 3,* Fifteenth Edition.
Hugh D. Young and Roger A. Freedman. Copyright © 2020 by Pearson Education, Inc. All rights reserved.

Practice Problem: What is the final kinetic energy of the electron if it has a velocity of 0.95c?
Answer: 1.13 MeV

Extra Practice: What is the total energy in this case? *Answer:* 1.64 MeV

5: Creating a particle

Two protons moving toward each other with equal speeds collide and produce an η_0 particle. (a) If the two protons and the η_0 are at rest after the collision, find the initial speed of the protons. (b) What is the kinetic energy of each proton? (c) What is the rest energy of the η_0 particle? The rest mass of each proton is 1.67×10^{-27} kg, and the rest mass of the η_0 is 9.75×10^{-28} kg.

IDENTIFY AND SET UP We'll use the relativistic energy relations to find the solution to the problem. We must include the rest energy of the particles.

EXECUTE Conservation of mass and energy requires that the energy before the interaction be the same as the energy after the interaction. After the interaction, there is only rest energy; before, there is only the total energy of the two protons. Energy–mass conservation gives

$$2\left(\frac{m_{proton}c^2}{\sqrt{1-v^2/c^2}}\right) = 2m_{proton}c^2 + m_{\eta_0}c^2.$$

We need to solve this equation for the velocity of the protons. Rearranging terms yields

$$\sqrt{1-v^2/c^2} = \frac{2m_{proton}}{2m_{proton}+m_{\eta_0}} = \frac{2(1.67\times10^{-27}\text{ kg})}{2(1.67\times10^{-27}\text{ kg})+(9.75\times10^{-28}\text{ kg})} = 0.774,$$

$$1-v^2/c^2 = (0.774)^2 = 0.5991,$$

$$v = \sqrt{(1-0.5991)}c = 0.633c.$$

The kinetic energy of each proton is

$$K = \frac{m_{proton}c^2}{\sqrt{1-v^2/c^2}} - m_{proton}c^2 = \frac{(1.67\times10^{-27}\text{ kg})c^2}{\sqrt{1-(0.633c)^2/c^2}} - (1.67\times10^{-27}\text{ kg})c^2$$

$$= 4.38\times10^{-11}\text{ J} = 274\text{ MeV}.$$

The rest mass of the η_0 particle is

$$E = m_{\eta_0}c^2 = (9.75\times10^{-28}\text{ kg})c^2 = 8.78\times10^{-11}\text{ J} = 548\text{ MeV}.$$

The initial speed of the protons is 0.663c. The initial kinetic energy of each proton is 274 MeV, and the rest energy of the η_0 particle is 548 MeV.

KEY CONCEPT 11 EVALUATE In this problem, we see how the kinetic energy of the two protons converts to the mass of the η_0 particle. Each proton loses 274 MeV, for a total of 548 MeV. This energy is converted into the rest energy of the η_0 particle.

Try It Yourself!

1: Traveling twin

One of a pair of twins makes a round-trip journey to a distant star, traveling at $0.95c$. If this twin ages 6.57 years during the journey, how much does his twin who remains on earth age during the trip? How far from earth did the twin travel?

Solution Checkpoints

IDENTIFY AND SET UP Use time dilation to solve the problem. Which twin measures proper time?

EXECUTE The time in the traveling twin's reference frame is the proper time, since he is the observer at rest. The twin on earth ages

$$\Delta t = \frac{\Delta t_0}{\sqrt{1 - u^2/c^2}} = 21.05 \text{ years,}$$

The distance traveled is the velocity times the elapsed time, given by

$$\Delta x = \frac{(0.95c)(21.05 \text{ years})}{c} = 20.0 \text{ light years.}$$

The total distance traveled is twice the maximum distance from earth. The traveling twin traveled 10 light years away from earth and back.

KEY CONCEPT 2 **EVALUATE** How can you check these results?

Practice Problem: How much would the traveling twin age if the spacecraft was traveling at $0.85c$? *Answer:* 12.4 years

Extra Practice: What is the twins' age difference if the spacecraft moved at $0.5c$ for the trip? *Answer:* 5.36 years

2: Traveling electrons

Two electrons traveling in the same direction have respective energies of 1.0 MeV and 2.0 MeV in reference frame S. Find the velocity of each electron in S. Find the velocity of the 2.0-MeV electron relative to the 1.0-MeV electron. Use 0.511 MeV for the rest energy of the electron.

Solution Checkpoints

IDENTIFY AND SET UP Use relativistic energy and velocity to solve the problem.

EXECUTE The velocity of each electron is found from

$$E = \left(\frac{m_{electron} c^2}{\sqrt{1 - u^2/c^2}} \right).$$

The 1.0-MeV electron has velocity $u_1 = 0.8602c$, and the 2.0 Me-V electron has velocity $u_2 = 0.9669c$. The relative velocities are found from

$$v'_x = \frac{v_x - u}{1 - u v_x/c^2}.$$

This equation is applied to a frame S' in which the 1.0-MeV electron is at rest. With this reference frame, v_x is the velocity of the 2.0-MeV electron in S (0.9669c), u is the velocity of S' with respect to S (0.8602c), and v'_x is the velocity of the 2.0-MeV electron in S' (the target variable). Substituting and solving leads to $v'_x = 0.636c$.

KEY CONCEPT 7 **EVALUATE** The challenge in this problem was to interpret the relative velocity variables carefully. With relativity, relative velocities are no longer simply vector sums of velocities.

Practice Problem: What is the velocity of the 1 MeV electron relative to the 2 MeV electron? *Answer:* –0.64c

Extra Practice: What is the relative speed if they are traveling in opposite directions? *Answer:* 0.997c

3: Neutron decay

When a neutron spontaneously decays into a proton, an electron, and a neutrino, the decay products are found to have a total kinetic energy of 7.81 MeV. The proton has a mass of 1.673×10^{-27} kg, the electron has a mass of 9.110×10^{-31} kg, and the neutrino has no mass. What is the mass of the neutron?

Solution Checkpoints

IDENTIFY AND SET UP Equate the energy of the reaction products to the rest energy of the neutron to solve the problem.

EXECUTE Energy conservation leads to

$$m_n c^2 = m_p c^2 + m_e c^2 + E_k.$$

Converting the kinetic energy to joules gives a kinetic energy of 1.25×10^{-13} J. Substituting values into the energy relation gives a neutron mass of 1.675×10^{-27} kg.

KEY CONCEPT 11 **EVALUATE** Can a proton spontaneously decay into a neutron?

Key Example Variation Problems

Solutions to these problems are in Chapter 37 of the Student's Solutions Manual.

Be sure to review EXAMPLES 37.1, 37.2, and 37.3 (Section 37.3) and EXAMPLES 37.4 and 37.5 (Section 37.4) before attempting these problems.

VP37.5.1 A positive K meson (K+) is an unstable subatomic particle that decays into other particles. At rest, its mean lifetime is 1.23×10^{-8} s. (a) What do you measure the mean lifetime of a K+ to be if it is moving at 0.800c relative to you? (b) As measured from the reference frame of this K+, what is the mean lifetime of a second K+ that is at rest relative to you?

VP37.5.2 Alia flies in her spacecraft at 0.600c relative to the planet Arrakis. As she passes Paul, at rest on Arrakis, they both start timers. (a) According to Alia's timer, 20.0 s elapses from when Paul starts his timer to when he stops his timer. What does Paul's timer read when he stops it? (b) Alia stops her timer when it reads 24.0 s.

From Chapter 37 of Student's Study Guide to accompany *University Physics with Modern Physics, Volume 3*, Fifteenth Edition. Hugh D. Young and Roger A. Freedman. Copyright © 2020 by Pearson Education, Inc. All rights reserved.

As measured by Paul, how much time elapses from when Alia starts her timer to when she stops it?

VP37.5.3 At one point in its orbit the earth is 1.50×10^{11} m from the sun. A spacecraft flies along a line from the earth to the sun at $0.950c$. (a) As measured by you on the earth, how much time does it take the spacecraft to travel the distance from the earth to the sun? (b) As measured by an astronaut on the spacecraft, what is the distance from the earth to the sun, and how much time does it take her to travel that distance?

VP37.5.4 A spacecraft flies past the earth in the direction toward the moon, a distance of 3.84×10^5 km. As measured by your clock on the earth, it takes 6.00 s for the clock on the spacecraft to tick off 2.00 s. You measure the length of the spacecraft in the direction of its motion to be 24.0 m. As measured by an astronaut at rest on the spacecraft, what are (a) the distance from the earth to the moon and (b) the length of the spacecraft?

Be sure to review EXAMPLES 37.6 and 37.7 (Section 37.5) before attempting these problems.

VP37.7.1 Gamora flies her spacecraft in the $+x$-direction past the planet Xandar at $0.750c$ while her sister Nebula is at rest on the surface of the planet. As Gamora passes Nebula, both sisters set their clocks to zero. Each chooses the zero of the x-axis to be at her position. (a) Nebula sets off fireworks next to her 2.50 s after Gamora passes her. What are the coordinates of this event as measured by Gamora?
(b) Gamora detects an explosion in space that occurs 4.00×10^8 m in front of her 2.50 s after she passes Nebula. What are the coordinates of this event as measured by Nebula?

VP37.7.2 Two events occur at different x-coordinates but at the same time as measured by Kamala. Doreen, who is moving at $0.800c$ relative to Kamala in the $+x$-direction, measures that one event takes place 0.600 s before the other. What is the distance between the two events as measured by (a) Kamala and (b) Doreen?

VP37.7.3 The spaceship *Nostromo* flies away from the earth in the $+x$-direction at $0.800c$. It fires a probe in the $-x$-direction at $0.600c$ relative to *Nostromo*. A scoutship is sent from the earth to recover the probe; the scoutship travels in the $+x$-direction at $0.700c$ relative to the earth. Find the velocity (in terms of c) of (a) the probe relative to the earth and (b) the scoutship relative to *Nostromo*.

VP37.7.4 As measured from the earth, the spaceship *Macross* is flying directly toward the earth at $0.750c$ while on the other side of the earth the spaceship *Yamato* is flying at $0.650c$ along the same line as *Macross*. Find the speed (in terms of c) of *Yamato* relative to *Macross* if *Yamato* is flying (a) toward the earth and (b) away from the earth.

Be sure to review EXAMPLE 37.9 (Section 37.7) and EXAMPLES 37.10 and 37.11 (Section 37.8) before attempting these problems.

VP37.11.1 A proton (rest mass 1.67×10^{-27} kg) is moving at $0.925c$. Find (a) the momentum of the proton, (b) its acceleration if a force of 9.00×10^{-14} N acts on the proton in the direction of its motion, and (c) its acceleration if a force of 9.00×10^{-14} N acts on the proton perpendicular to the direction of its motion.

VP37.11.2 A moving electron (rest mass 9.11×10^{-31} kg) has total energy 4.00×10^{-13} J. Find (a) the kinetic energy of the electron, (b) its Lorentz factor γ, and (c) its speed in terms of c.

VP37.11.3 A moving electron (rest mass 9.11×10^{-31} kg) has momentum 2.60×10^{-22} kg·m/s. Find (a) the total energy of the electron, (b) its Lorentz factor γ, and (c) its speed in terms of c.

VP37.11.4 The $\psi(2S)$ meson is an unstable particle with rest energy 3686 MeV. One way in which a $\psi(2S)$ can decay is into a positive K meson (K+) and a negative K meson (K$^-$), each with rest energy 495 MeV. Assume the $\psi(2S)$ is at rest before it decays. Find (a) the kinetic energy in MeV, (b) the Lorentz factor γ, and (c) the speed (in terms of c) of each K meson.

STUDENT'S SOLUTIONS MANUAL FOR RELATIVITY

VP37.5.1. **IDENTIFY:** This problem is about time dilation.

SET UP: $\Delta t = \dfrac{\Delta t_0}{\sqrt{1 - u^2/c^2}}$. We want the mean lifetime of the K^+ particle.

EXECUTE: **(a)** At $u = 0.800c$: $\Delta t = \dfrac{1.23 \times 10^{-8} \text{ s}}{\sqrt{1 - (0.800)^2}} = 2.05 \times 10^{-8}$ s.

(b) At $u = 0.800\,c$: To the first K^+, the second one is moving at $u = 0.800c$, so the answer is the same as in part (a): $\Delta t = 2.05 \times 10^{-8}$ s.

EVALUATE: The relative speed of the K^+ is the same in both cases, so the lifetime is the same.

VP37.5.2. **IDENTIFY:** This problem is about time dilation.

SET UP: $\Delta t = \dfrac{\Delta t_0}{\sqrt{1 - u^2/c^2}}$. We want the time interval that both Paul and Alia read.

EXECUTE: **(a)** Paul's timer reads the proper time, so solve for Δt_0.

$\Delta t_0 = \Delta t \sqrt{1 - u^2/c^2} = (20.0 \text{ s})\sqrt{1 - (0.600)^2} = 16.0$ s.

(b) Alia's time is the proper time. $\Delta t = \dfrac{24.0 \text{ s}}{\sqrt{1 - (0.600)^2}} = 30.0$ s.

EVALUATE: The proper time is the time measured by the observer at rest in her frame.

VP37.5.3. **IDENTIFY:** This problem is about length contraction.

SET UP: $L = L_0 \sqrt{1 - u^2/c^2}$. We want the distances and times measured.

EXECUTE: **(a)** In your frame, the proper length is 1.50×10^{11} m.

$\Delta t = d/u = (1.50 \times 10^{11} \text{ m})/(0.950c) = 526$ s.

(b) For the astronaut, the distance is length contracted.

$L = (1.50 \times 10^{11} \text{ m})\sqrt{1 - (0.950)^2} = 4.68 \times 10^{10}$ m. For the astronaut, the sun is moving toward him

at $0.950c$, so $\Delta t = L/u = (4.68 \times 10^{10} \text{ m})/(0.950c) = 164$ s.

EVALUATE: Check: Use time dilation. To a person on Earth, the trip takes 526 s. The proper time

is the astronaut's time, so $\Delta t_E = \dfrac{\Delta t_A}{\sqrt{1 - (0.950)^2}} = 526$ s.

VP37.5.4. **IDENTIFY:** This problem is about length contraction.

SET UP: $L = L_0 \sqrt{1 - u^2/c^2}$, $\Delta t = \dfrac{\Delta t_0}{\sqrt{1 - u^2/c^2}}$. For you, it takes 6.00 s for the rocket clock to read 2.00 s. You measure $L_{rocket} = 24.0$ m. We want the distances measured by the astronaut in the spacecraft.

EXECUTE: **(a)** Earth-Moon distance: First find the spacecraft's speed relative to Earth. The proper time is 2.00 s in the rocket. $6.00 \text{ s} = \dfrac{2.00 \text{ s}}{\sqrt{1 - u^2/c^2}}$, so $\sqrt{1 - u^2/c^2} = 0.333$. To the astronaut, the Earth-Moon distance is length-contracted. $L = L_0 \sqrt{1 - u^2/c^2} = (3.84 \times 10^5 \text{ km})(0.333) = 1.28 \times 10^5$ km.

(b) The proper length of the spacecraft is measured by the astronaut, so $L_0 = \dfrac{L}{\sqrt{1 - u^2/c^2}} = \dfrac{24.0 \text{ m}}{0.333}$ = 72.0 m.

EVALUATE: The Earth observer sees the spacecraft length-contracted and the astronaut sees the Earth-Moon distance length-contracted.

VP37.7.1. **IDENTIFY:** This problem is about the Lorentz transformation equations.

SET UP: $x' = \gamma(x - ut)$, $t' = \gamma\left(t - xu/c^2\right)$. $\gamma = \dfrac{1}{\sqrt{1 - u^2/c^2}} = \dfrac{1}{\sqrt{1 - (0.750)^2}} = 1.512$. Gamora is the primed frame and Nebula is the unprimed frame.

EXECUTE: **(a)** We want the coordinates as measured by Gamora. Using $x' = \gamma(x - ut)$, we have $ut = (0.750c)(2.50 \text{ s}) = 5.625 \times 10^8$ m. $x' = (1.512)(0 - 5.625 \times 10^8 \text{ m}) = 8.50 \times 10^8$ m. Using $t' = \gamma\left(t - xu/c^2\right)$ gives $t' = (1.512)(2.50 \text{ s} - 0) = 3.78$ s.

(b) We want the coordinates as measured by Nebula. She is the unprimed frame, so $x = \gamma(x' + ut)$ and $t = \gamma\left(t' + x'u/c^2\right)$. $x'u/c^2 = (4.00 \times 10^8 \text{ m})(0.750c)/c^2 = 1.00$ s. This gives $x = \gamma(x' + ut) = (1.512)(4.00 \times 10^8 \text{ m} + 5.625 \times 10^8 \text{ m}) = 1.46 \times 10^9$ m. $t = \gamma\left(t' + x'u/c^2\right) = (1.512)(2.50 \text{ s} + 1.00 \text{ s}) = 5.29$ s.

EVALUATE: Be very careful to decide which are the primed and unprimed frames.

VP37.7.2. **IDENTIFY:** This problem is about the Lorentz transformation equations.

SET UP: $x' = \gamma(x - ut)$, $t' = \gamma\left(t - xu/c^2\right)$. $\gamma = \dfrac{1}{\sqrt{1 - u^2/c^2}} = \dfrac{1}{\sqrt{1 - (0.800)^2}} = 1.6667$. Doreen is the primed frame and Kamala is the unprimed frame. In Kamala's frame, the events are simultaneous but occur at different places. In Doreen's frame the events are 0.600 s apart. We want the distance between the two events as measured by Doreen and Kamala.

EXECUTE: **(a)** In Kamala's frame: Use the Lorentz transformation equation for time and solve for Δx_K. $\Delta t'_D = \gamma\left(\Delta t_K - x_K u/c^2\right)$. $0.600 \text{ s} = (1.6667)\left(0 - \Delta x_K(0.800c)/c^2\right)$. $\Delta x_K = 1.35 \times 10^8$ m.

(b) In Doreen's frame: $\Delta x'_D = \gamma(\Delta x_K - u\Delta t_K) = \Delta x'_D = \gamma(\Delta x_K - 0) = (1.6667)(1.35 \times 10^8 \text{ m}) = 2.25 \times 10^8$ m.

EVALUATE: The distances are different because of their relative motion.

VP37.7.3. **IDENTIFY:** This problem is about relative velocity.

SET UP: $v'_x = \dfrac{v_x - u}{1 - uv_x/c^2}$, $v_x = \dfrac{v'_x + u}{1 + uv'_x/c^2}$. Fig. 37.7.3 shows the motions involved.

Scout   0.700c (relative to earth)   Probe ← 0.600c (relative to Nostromo)   Nostromo → 0.800c (relative to earth)

Figure VP37.7.3

EXECUTE: **(a)** We want the velocity of the probe relative to Earth. Let *Nostromo* be the primed frame moving relative to Earth at $0.800c$ in the $+x$ direction. Earth is the unprimed frame.

$$v_x = \frac{v'_x + u}{1 + uv'_x/c^2} = \frac{-0.600c + 0.800c}{1 + (0.800c)(-0.600c)/c^2} = +0.385c.$$

(b) We want the velocity of the Scout ship relative to *Nostromo*. Let *Nostromo* be the primed frame and Earth the unprimed frame. We want v'_x. Using $v'_x = \dfrac{v_x - u}{1 - uv_x/c^2}$ gives

$$v'_x = \frac{0.700c - 0.800c}{1 - (0.800c)(0.700c)/c^2} = -0.227c.$$

EVALUATE: The result in (b) is reasonable because *Nostromo* is moving faster than the Scout probe, so relative to *Nostromo*, the Scout is moving in the $-x$ direction.

VP37.7.4. **IDENTIFY** This problem is about relative velocity.

SET UP: $v'_x = \dfrac{v_x - u}{1 - uv_x/c^2}$. We want the speed of *Yamato* relative to *Macross*.

EXECUTE: **(a)** *Yamato* is flying toward Earth. Let *Macross* be the primed frame and Earth the unprimed frame. We want v'_Y. $v'_x = \dfrac{v_x - u}{1 - uv_x/c^2}$ becomes $v'_Y = \dfrac{v_Y - u}{1 - uv_Y/c^2}$

$$= \frac{-650c - 0.750c}{1 - (0.750c)(-0.650c)/c^2} = -0.941c. \text{ } Yamato\text{'s speed is } 0.941c.$$

(b) *Yamato* is flying away from Earth. $v'_Y = \dfrac{+650c - 0.750c}{1 - (0.750c)(+0.650c)/c^2} = -0.195c.$

EVALUATE: In both cases *Yamato* has a negative velocity component because relative to *Macross*, *Yamato* is getting closer, hence moving in the $-x$ direction. The relative speed in (b) is less than that in (a), as expected.

VP37.11.1. **IDENTIFY:** This problem deals with the force on a proton and its momentum at high speeds.

SET UP and **EXECUTE:** $\gamma = \dfrac{1}{\sqrt{1 - u^2/c^2}} = \dfrac{1}{\sqrt{1 - (0.950)^2}} = 3.20256.$ **(a)** We want the momentum.

$p = m\gamma c = (1.67 \times 10^{-27} \text{ kg})(3.20256)c = 1.52 \times 10^{-18} \text{ kg} \cdot \text{m/s}.$

(b) We want the acceleration. If the force and velocity are along the same line, $F = \gamma^3 ma$. So

$a = \dfrac{F}{m\gamma^3}$. Using the given numbers gives $a = 1.64 \times 10^{12} \text{ m/s}^2$.

(c) We want the acceleration. If the force and velocity are perpendicular, $F = \gamma ma$. So $a = \dfrac{F}{m\gamma}$.

Using the given numbers we get $a = 1.68\times10^{13}$ m/s^2.

EVALUATE: Without relativity the acceleration would be $a = F/m = 5.39\times10^{13}$ m/s^2, which is *greater* than when relativity it taken into account.

VP37.11.2. **IDENTIFY:** This problem is about the energy of a moving electron.
SET UP and **EXECUTE:** **(a)** We want the kinetic energy. $E = K + mc^2$, so $K = E - mc^2$
$= 4.00\times10^{-13}$ J $+ (9.11\times10^{-31}$ kg)$c^2 = 3.18\times10^{-13}$ J.

(b) We want γ. $E = m\gamma c^2$. Solve for γ and use the given numbers. $\gamma = \dfrac{E}{mc^2} = 4.88$.

(c) We want the speed. Solve $\gamma = \dfrac{1}{\sqrt{1-v^2/c^2}}$ for v. $v/c = \sqrt{1-1/\gamma^2} = \sqrt{1-1/4.88^2} = 0.979$, so $v = 0.979c$.

EVALUATE: At speeds near the speed of light, the kinetic energy is much different from $\dfrac{1}{2}mv^2$.

VP37.11.3. **IDENTIFY:** This problem is about the energy of a moving electron.
SET UP: $E^2 = (pc)^2 + (mc^2)^2$, $E = m\gamma c^2$, $v/c = \sqrt{1-1/\gamma^2}$.

EXECUTE: **(a)** We want the total energy E. Solve $E^2 = (pc)^2 + (mc^2)^2$ for E and use the given numbers for the momentum p and rest mass m. The result is $E = 1.13\times10^{-13}$ J.

(b) We want γ. Solve $E = m\gamma c^2$ for γ and use the given m and E we found in (a), giving $\gamma = 1.38$.

(c) We want the speed v. $v/c = \sqrt{1-1/\gamma^2} = \sqrt{1-1/1.38^2} = 0.689$, so $v = 0.689c$.

EVALUATE: Careful! $E \neq pc + mc^2$.

VP37.11.4 **IDENTIFY:** This problem is about energy in particle decay.
SET UP: The $\psi(2S)$ is at rest, so by momentum conservation the K mesons have equal but opposite momentum and therefore equal speeds and equal kinetic energies. $mc^2 = 495$ MeV for each K meson. $E = K + mc^2$, $K = (\gamma-1)mc^2$.

EXECUTE: **(a)** We want the kinetic energy K of each meson. Using $E = K + mc^2$ gives 3686 MeV $= (K + 495$ MeV$) + (K + 495$ MeV$)$. Solving gives $K = 1348$ MeV for each meson.

(b) We want γ. $K = (\gamma-1)mc^2 = (\gamma-1)(495$ MeV$)$. Solve for γ and use $K = 1348$ MeV from part (a), giving $\gamma = 3.72$.

(c) We want the speed. $v/c = \sqrt{1-1/\gamma^2} = \sqrt{1-1/3.72^2} = 0.963$, so $v = 0.963c$.

EVALUATE: The K$^+$ and K$^-$ mesons are highly relativistic with $v = 0.963c$ and $\gamma = 3.72$.

37.1. **IDENTIFY** and **SET UP:** Consider the distance A to O' and B to O' as observed by an observer on the ground (Figure 37.1).

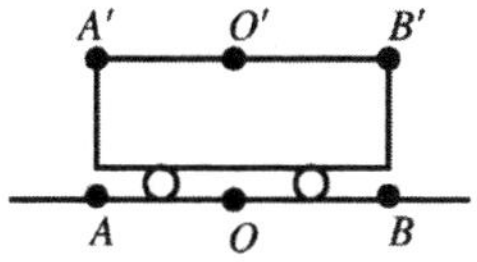

Figure 37.1

EXECUTE: The statement that the events are simultaneous to an observer on the train means that light pulses from A' and B' arrive at O' at the same time. To the observer at O, light from A' has a longer distance to travel than light from B' so O will conclude that the pulse from $A(A')$ started before the pulse at $B(B')$. To the observer at O, bolt A appeared to strike first.

EVALUATE: Section 37.2 shows that if the events are simultaneous to the observer on the ground, then an observer on the train measures that the bolt at B' struck first.

37.5. **(a) IDENTIFY** and **SET UP:** $\Delta t_0 = 2.60 \times 10^{-8}$ s; $\Delta t = 4.20 \times 10^{-7}$ s. In the lab frame the pion is created and decays at different points, so this time is not the proper time.

EXECUTE: $\Delta t = \dfrac{\Delta t_0}{\sqrt{1 - u^2/c^2}}$ says $1 - \dfrac{u^2}{c^2} = \left(\dfrac{\Delta t_0}{\Delta t}\right)^2$.

$\dfrac{u}{c} = \sqrt{1 - \left(\dfrac{\Delta t_0}{\Delta t}\right)^2} = \sqrt{1 - \left(\dfrac{2.60 \times 10^{-8}\ \text{s}}{4.20 \times 10^{-7}\ \text{s}}\right)^2} = 0.998; \ u = 0.998c.$

EVALUATE: $u < c$, as it must be, but u/c is close to unity and the time dilation effects are large.

(b) IDENTIFY and **SET UP:** The speed in the laboratory frame is $u = 0.998c$; the time measured in this frame is Δt, so the distance as measured in this frame is $d = u\Delta t$.

EXECUTE: $d = (0.998)(2.998 \times 10^8\ \text{m/s})(4.20 \times 10^{-7}\ \text{s}) = 126$ m.

EVALUATE: The distance measured in the pion's frame will be different because the time measured in the pion's frame is different (shorter).

37.11. **IDENTIFY** and **SET UP:** The 2.2 μs lifetime is Δt_0 and the observer on earth measures Δt. The atmosphere is moving relative to the muon so in its frame the height of the atmosphere is l and l_0 is 10 km.

EXECUTE: **(a)** The greatest speed the muon can have is c, so the greatest distance it can travel in 2.2×10^{-6} s is $d = vt = (3.00 \times 10^8\ \text{m/s})(2.2 \times 10^{-6}\ \text{s}) = 660\ \text{m} = 0.66$ km.

(b) $\Delta t = \dfrac{\Delta t_0}{\sqrt{1 - u^2/c^2}} = \dfrac{2.2 \times 10^{-6}\ \text{s}}{\sqrt{1 - (0.999)^2}} = 4.9 \times 10^{-5}$ s.

$d = vt = (0.999)(3.00 \times 10^8\ \text{m/s})(4.9 \times 10^{-5}\ \text{s}) = 15$ km.

In the frame of the earth the muon can travel 15 km in the atmosphere during its lifetime.

(c) $l = l_0\sqrt{1 - u^2/c^2} = (10\ \text{km})\sqrt{1 - (0.999)^2} = 0.45$ km.

EVALUATE: In the frame of the muon the height of the atmosphere is less than the distance it moves during its lifetime.

37.13. **IDENTIFY:** Apply $l = l_0 \sqrt{1 - u^2/c^2}$.

SET UP: The proper length l_0 of the runway is its length measured in the Earth's frame. The proper time Δt_0 for the time interval for the spacecraft to travel from one end of the runway to the other is the time interval measured in the frame of the spacecraft.

EXECUTE: **(a)** $l_0 = 3600$ m.

$$l = l_0 \sqrt{1 - \frac{u^2}{c^2}} = (3600 \text{ m}) \sqrt{1 - \frac{(4.00 \times 10^7 \text{ m/s})^2}{(3.00 \times 10^8 \text{ m/s})^2}} = (3600 \text{ m})(0.991) = 3568 \text{ m}.$$

(b) $\Delta t = \dfrac{l_0}{u} = \dfrac{3600 \text{ m}}{4.00 \times 10^7 \text{ m/s}} = 9.00 \times 10^{-5}$ s.

(c) $\Delta t_0 = \dfrac{l}{u} = \dfrac{3568 \text{ m}}{4.00 \times 10^7 \text{ m/s}} = 8.92 \times 10^{-5}$ s.

EVALUATE: $\dfrac{1}{\gamma} = 0.991$, so $\Delta t = \gamma \Delta t_0$ gives $\Delta t = \dfrac{8.92 \times 10^{-5} \text{ s}}{0.991} = 9.00 \times 10^{-5}$ s. The result from length contraction is consistent with the result from time dilation.

37.17. **IDENTIFY:** The relativistic velocity addition formulas apply since the speeds are close to that of light.

SET UP: The relativistic velocity addition formula is $v_x' = \dfrac{v_x - u}{1 - \dfrac{u v_x}{c^2}}$.

EXECUTE: **(a)** For the pursuit ship to catch the cruiser, the distance between them must be decreasing, so the velocity of the cruiser relative to the pursuit ship must be directed toward the pursuit ship.

(b) Let the unprimed frame be Tatooine and let the primed frame be the pursuit ship. We want the velocity v' of the cruiser knowing the velocity of the primed frame u and the velocity of the cruiser v in the unprimed frame (Tatooine). $v_x' = \dfrac{v_x - u}{1 - \dfrac{u v_x}{c^2}} = \dfrac{0.600c - 0.800c}{1 - (0.600)(0.800)} = -0.385c.$

The result implies that the cruiser is moving toward the pursuit ship at $0.385c$.

EVALUATE: The nonrelativistic formula would have given $-0.200c$, which is considerably different from the correct result.

37.19. **IDENTIFY** and **SET UP:** Reference frames S and S' are shown in Figure 37.19.

Frame S is at rest in the laboratory. Frame S' is attached to particle 1.

Figure 37.19

u is the speed of S' relative to S; this is the speed of particle 1 as measured in the laboratory. Thus $u = +0.650c$. The speed of particle 2 in S' is $0.950c$. Also, since the two particles move in opposite directions, 2 moves in the $-x'$-direction and $v'_x = -0.950c$. We want to calculate v_x, the speed of particle 2 in frame S, so use $v_x = \dfrac{v'_x + u}{1 + uv'_x/c^2}$.

EXECUTE: $v_x = \dfrac{v'_x + u}{1 + uv'_x/c^2} = \dfrac{-0.950c + 0.650c}{1 + (0.650c)(-0.950c)/c^2} = \dfrac{-0.300c}{1 - 0.6175} = -0.784c$. The speed of the second particle, as measured in the laboratory, is $0.784c$.

EVALUATE: The incorrect Galilean expression for the relative velocity gives that the speed of the second particle in the lab frame is $0.300c$. The correct relativistic calculation gives a result more than twice this.

37.21. **IDENTIFY:** The relativistic velocity addition formulas apply since the speeds are close to that of light.

SET UP: The relativistic velocity addition formula is $v'_x = \dfrac{v_x - u}{1 - \dfrac{uv_x}{c^2}}$.

EXECUTE: In the relativistic velocity addition formula for this case, v'_x is the relative speed of particle 1 with respect to particle 2, v is the speed of particle 2 measured in the laboratory, and u is the speed of particle 1 measured in the laboratory, $u = -v$.

$v'_x = \dfrac{v - (-v)}{1 - (-v)v/c^2} = \dfrac{2v}{1 + v^2/c^2}$. $\dfrac{v'_x}{c^2}v^2 - 2v + v'_x = 0$ and $(0.890c)v^2 - 2c^2v + (0.890c^3) = 0$.

This is a quadratic equation with solution $v = 0.611c$ (v must be less than c).

EVALUATE: The nonrelativistic result would be $0.445c$, which is considerably different from this result.

37.23. **IDENTIFY** and **SET UP:** Source and observer are approaching, so use $f = \sqrt{\dfrac{c+u}{c-u}}f_0$. Solve for u, the speed of the light source relative to the observer.

EXECUTE: **(a)** $f^2 = \left(\dfrac{c+u}{c-u}\right)f_0^2$.

$(c-u)f^2 = (c+u)f_0^2$ and $u = \dfrac{c(f^2 - f_0^2)}{f^2 + f_0^2} = c\left(\dfrac{(f/f_0)^2 - 1}{(f/f_0)^2 + 1}\right)$.

$\lambda_0 = 675$ nm, $\lambda = 575$ nm.

$u = \left(\dfrac{(675 \text{ nm}/575 \text{ nm})^2 - 1}{(675 \text{ nm}/575 \text{ nm})^2 + 1}\right)c = 0.159c = (0.159)(2.998\times10^8 \text{ m/s}) = 4.77\times10^7 \text{ m/s}$; definitely speeding

(b) $4.77\times10^7 \text{ m/s} = (4.77\times10^7 \text{ m/s})(1 \text{ km}/1000 \text{ m})(3600 \text{ s}/1 \text{ h}) = 1.72\times10^8 \text{ km/h}$. Your fine would be $\$1.72\times10^8$ (172 million dollars).

EVALUATE: The source and observer are approaching, so $f > f_0$ and $\lambda < \lambda_0$. Our result gives $u < c$, as it must.

37.25. **IDENTIFY:** The problem involves momentum and the Lorentz factor.

SET UP: $p = m\gamma v$, $\gamma = \dfrac{1}{\sqrt{1 - u^2/c^2}}$, $u/c = \sqrt{1 - 1/\gamma^2}$, particle 1: $\gamma_1 = 1.12$, particle 2: $u_2 = 2u_1$.

EXECUTE: (a) We want γ_2. First find u_1. $u_1 = c\sqrt{1-1/\gamma^2} = c\sqrt{1-1/1.12^2} = 0.450343c$, so $u_2 = 2u_1$

$= 0.90068c$. $\gamma_2 = \dfrac{1}{\sqrt{1-(0.90068)^2}} = 2.30$.

(b) We want p_2/p_1. $\dfrac{p_2}{p_1} = \dfrac{m\gamma_2 u_2}{m\gamma_1 u_1} = \dfrac{\gamma_2(2u_1)}{\gamma_1 u_1} = \dfrac{2\gamma_2}{\gamma_1} = \dfrac{2(2.30)}{1.12} = 4.11$.

EVALUATE: For speeds much less than that of light, we would have $p_2 = 2p_1$, but near the speed of light we find $p_2 = 4.11p_1$, a very significant difference.

37.29. IDENTIFY: Apply $p = \dfrac{mv}{\sqrt{1-v^2/c^2}}$ and $F = \gamma^3 ma$.

SET UP: For a particle at rest (or with $v \ll c$), $a = F/m$.

EXECUTE: (a) $p = \dfrac{mv}{\sqrt{1-v^2/c^2}} = 2mv$.

$\Rightarrow 1 = 2\sqrt{1-v^2/c^2} \Rightarrow \dfrac{1}{4} = 1 - \dfrac{v^2}{c^2} \Rightarrow v^2 = \dfrac{3}{4}c^2 \Rightarrow v = \dfrac{\sqrt{3}}{2}c = 0.866c$.

(b) $F = \gamma^3 ma = 2ma \Rightarrow \gamma^3 = 2 \Rightarrow \gamma = (2)^{1/3}$ so $\dfrac{1}{1-v^2/c^2} = 2^{2/3} \Rightarrow \dfrac{v}{c} = \sqrt{1-2^{-2/3}} = 0.608$.

EVALUATE: The momentum of a particle and the force required to give it a given acceleration both increase without bound as the speed of the particle approaches c.

37.31. IDENTIFY: Apply $K = \dfrac{mc^2}{\sqrt{1-v^2/c^2}} - mc^2$.

SET UP: The rest energy is mc^2.

EXECUTE: (a) $K = \dfrac{mc^2}{\sqrt{1-v^2/c^2}} - mc^2 = mc^2$.

$\Rightarrow \dfrac{1}{\sqrt{1-v^2/c^2}} = 2 \Rightarrow \dfrac{1}{4} = 1 - \dfrac{v^2}{c^2} \Rightarrow v = \sqrt{\dfrac{3}{4}}c = 0.866c$.

(b) $K = 5mc^2 \Rightarrow \dfrac{1}{\sqrt{1-v^2/c^2}} = 6 \Rightarrow \dfrac{1}{36} = 1 - \dfrac{v^2}{c^2} \Rightarrow v = \sqrt{\dfrac{35}{36}}c = 0.986c$.

EVALUATE: If $v \ll c$, then K is much less than the rest energy of the particle.

37.35. IDENTIFY and SET UP: The total energy is given in terms of the momentum by $E^2 = (mc^2)^2 + (pc)^2$. In terms of the total energy E, the kinetic energy K is $K = E - mc^2$. The rest energy is mc^2.

EXECUTE: (a)

$E = \sqrt{(mc^2)^2 + (pc)^2} = \sqrt{\left[(6.64\times10^{-27})(2.998\times10^8)^2\right]^2 + \left[(2.10\times10^{-18})(2.998\times10^8)\right]^2}$ J.

$E = 8.67\times10^{-10}$ J.

(b) $mc^2 = (6.64\times10^{-27}$ kg$)(2.998\times10^8$ m/s$)^2 = 5.97\times10^{-10}$ J.

$K = E - mc^2 = 8.67\times10^{-10}$ J $- 5.97\times10^{-10}$ J $= 2.70\times10^{-10}$ J.

(c) $\dfrac{K}{mc^2} = \dfrac{2.70\times10^{-10}\text{ J}}{5.97\times10^{-10}\text{ J}} = 0.452.$

EVALUATE: The incorrect nonrelativistic expressions for K and p give $K = p^2/2m = 3.3\times10^{-10}$ J; the correct relativistic value is less than this.

37.39. **IDENTIFY and SET UP:** Use $K = q\Delta V = e\Delta V$ and conservation of energy to relate the potential difference to the kinetic energy gained by the electron. Use $K = mc^2\left(\dfrac{1}{\sqrt{1-v^2/c^2}} - 1\right)$ to calculate the kinetic energy from the speed.

EXECUTE: (a) $K = q\Delta V = e\Delta V.$

$K = mc^2\left(\dfrac{1}{\sqrt{1-v^2/c^2}} - 1\right) = 4.025mc^2 = 3.295\times10^{-13}$ J $= 2.06$ MeV.

$\Delta V = K/e = 2.06\times10^6$ V.

(b) From part (a), $K = 3.30\times10^{-13}$ J $= 2.06$ MeV.

EVALUATE: The speed is close to c and the kinetic energy is four times the rest mass.

37.41. **IDENTIFY and SET UP:** There must be a length contraction such that the length a becomes the same as b; $l_0 = a$, $l = b$. l_0 is the distance measured by an observer at rest relative to the spacecraft. Use $l = l_0\sqrt{1-u^2/c^2}$ and solve for u.

EXECUTE: $\dfrac{l}{l_0} = \sqrt{1-u^2/c^2}$ so $\dfrac{b}{a} = \sqrt{1-u^2/c^2}$;

$a = 1.40b$ gives $b/1.40b = \sqrt{1-u^2/c^2}$. and thus $1-u^2/c^2 = 1/(1.40)^2$.

$u = \sqrt{1-1/(1.40)^2}\,c = 0.700c = 2.10\times10^8$ m/s.

EVALUATE: A length on the spacecraft in the direction of the motion is shortened. A length perpendicular to the motion is unchanged.

37.43. **IDENTIFY and SET UP:** The proper time Δt_0 is the time that elapses in the frame of the space probe. Δt is the time that elapses in the frame of the earth. The distance traveled is 42.2 light years, as measured in the earth frame.

EXECUTE: Light travels 42.2 light years in 42.2 y, so $\Delta t = \left(\dfrac{c}{0.9930c}\right)(42.2\text{ y}) = 42.5$ y.

$\Delta t_0 = \Delta t\sqrt{1-u^2/c^2} = (42.5\text{ y})\sqrt{1-(0.9930)^2} = 5.0$ y. She measures her biological age to be $19\text{ y} + 5.0\text{ y} = 24.0$ y.

EVALUATE: Her age measured by someone on earth is $19\text{ y} + 42.5\text{ y} = 61.5$ y.

37.45. **IDENTIFY:** This problem involves relative velocity and mass.

SET UP: $m_{rel} = m\gamma = \dfrac{m}{\sqrt{1-u^2/c^2}}$, $v_x' = \dfrac{v_x - u}{1-uv_x/c^2}$. We want the mass of the standard kilogram.

EXECUTE: (a) In classical physics, mass doesn't change with speed, so both crews would measure 1.0 kg.

(b) <u>For A</u>: $m_{\text{rel}} = m\gamma_A = \dfrac{m}{\sqrt{1 - u_A^2/c^2}} = \dfrac{1.00 \text{ kg}}{\sqrt{1 - 0.80^2}} = 1.7 \text{ kg}.$

<u>For B</u>: We need to find the speed of B relative to Earth. Use $v_x' = \dfrac{v_x - u}{1 - uv_x/c^2}$, letting Earth be the primed frame and A the unprimed frame. We want v'. $v_{B/E}' = \dfrac{0.98c - 0.80c}{1 - (0.98c)(0.80c)/c^2} = 0.833c.$ The

mass is $m_{\text{rel}} = \dfrac{m}{\sqrt{1 - u_B^2/c^2}} = \dfrac{1.00 \text{ kg}}{\sqrt{1 - 0.833^2}} = 1.8 \text{ kg}.$

EVALUATE: B measures a greater mass than A measures because B is moving faster relative to Earth than A is. Both of them, however, would measure 1.00 kg as the rest mass of the standard kilogram.

37.47. **IDENTIFY:** Since the speed is very close to the speed of light, we must use the relativistic formula for kinetic energy.

SET UP: The relativistic formula for kinetic energy is $K = mc^2 \left(\dfrac{1}{\sqrt{1 - v^2/c^2}} - 1 \right)$ and the relativistic mass is $m_{\text{rel}} = \dfrac{m}{\sqrt{1 - v^2/c^2}}.$

EXECUTE: **(a)** $K = 7.0 \times 10^{12} \text{ eV} = 1.12 \times 10^{-6} \text{ J}.$ Using this value in the relativistic kinetic energy formula and substituting the mass of the proton for m, we get $K = mc^2 \left(\dfrac{1}{\sqrt{1 - v^2/c^2}} - 1 \right)$ which

gives $\dfrac{1}{\sqrt{1 - v^2/c^2}} = 7.45 \times 10^3$ and $1 - \dfrac{v^2}{c^2} = \dfrac{1}{(7.45 \times 10^3)^2}.$ Solving for v gives

$1 - \dfrac{v^2}{c^2} = \dfrac{(c + v)(c - v)}{c^2} = \dfrac{2(c - v)}{c}$, since $c + v \approx 2c.$ Substituting $v = (1 - \Delta)c$, we have

$1 - \dfrac{v^2}{c^2} = \dfrac{2(c - v)}{c} = \dfrac{2[c - (1 - \Delta)c]}{c} = 2\Delta.$ Solving for Δ gives

$\Delta = \dfrac{1 - v^2/c^2}{2} = \dfrac{\dfrac{1}{(7.45 \times 10^3)^2}}{2} = 9.0 \times 10^{-9}.$

(b) Using the relativistic mass formula and the result that $\dfrac{1}{\sqrt{1 - v^2/c^2}} = 7.45 \times 10^3$, we have

$m_{\text{rel}} = \dfrac{m}{\sqrt{1 - v^2/c^2}} = m \left(\dfrac{1}{\sqrt{1 - v^2/c^2}} \right) = (7.5 \times 10^3)m.$

EVALUATE: At such high speeds, the proton's mass is 7500 times as great as its rest mass.

$$a = \frac{F}{m\gamma^3} = \frac{F}{m}\left(1 - \frac{v^2}{c^2}\right)^{3/2} \rightarrow a^2 = \left(\frac{F}{m}\right)^2\left(1 - \frac{v^2}{c^2}\right)^3.$$

From this result we see that a graph of a^2 versus $\left(1 - \frac{v^2}{c^2}\right)^3$ should be a straight line with slope equal to $(F/m)^2$. Figure 37.65 shows the graph of the data in the table with the problem. It is well fit by a straight line having slope equal to 1.608×10^9 m^2/s^4. Therefore the mass is

$$(F/m)^2 = \text{slope} \rightarrow m = \frac{F}{\sqrt{\text{slope}}} = \frac{8.00 \times 10^{-14}\text{ N}}{\sqrt{1.608 \times 10^9\text{ m}^2/\text{s}^4}} = 2.0 \times 10^{-18}\text{ kg}.$$

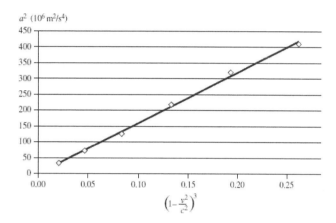

Figure 37.65

(b) In this case, $v \ll c$, so γ is essentially equal to 1. Therefore we can use the familiar form of Newton's second law, $F = ma$.
$a = F/m = (8.00 \times 10^{-14}\text{ N})/(2.0 \times 10^{-18}\text{ kg}) = 4.0 \times 10^4$ m/s^2.
EVALUATE: When v is close to c, $a = F/m$ does not give the correct result. For example, using data from the table in the problem, when $v/c = 0.85$, the acceleration is measured to be 5900 m/s^2. But using the familiar Newtonian formula we get $a = F/m = (8.00 \times 10^{-14}\text{ N})/(2.0 \times 10^{-18}\text{ kg}) = 4.0 \times 10^4$ m/s^2 = 40,000 m/s^2, which is *very* different from the relativistic result of 5900 m/s^2.

37.67. **IDENTIFY:** We are using the Lorentz transformation equations to investigate space-time diagrams.

SET UP: S is the reference frame of the train station and S' is the moving frame of the rocket train. $x' = \gamma(x - vt)$, $t' = \gamma\left(t - xv/c^2\right)$.

EXECUTE: **(a)** Event 1: $x_1 = 0$, $t_1 = 0$, so $x_1' = 0$, $t_1' = 0$. So $(x_1', t_1') = (0, 0)$.

Event 2: $x_2 = -L$, $t_2 = 0$. $x_2' = -L\gamma$, $t_2' = [-(-L)v/c^2]\gamma = Lv\gamma/c^2$. So $(x_2', t_2') = (-L\gamma, Lv\gamma/c^2)$.

Event 3: $x_3 = 0$, $t_3 = T$. $x_3' = -vT\gamma$, $t_3' = T\gamma$. So $(x_3', t_3') = (-vT\gamma, T\gamma)$.

Event 4: $x_4 = -L$, $t_4 = T$. $x_4' = -(L + vT)\gamma$, $t_4' = (T + Lv/c^2)\gamma$. So $(x_4', t_4') = \left((-L + vT)\gamma, (T + Lv/c^2)\gamma\right)$.

(b) Fig. 37.67 shows the spacetime diagram.

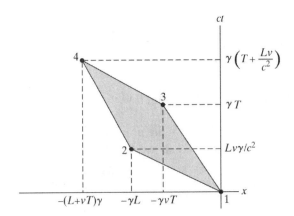

Figure 37.67

(c) We want the area in frame S. $A = LcT = cLT$.

(d) We want the area in frame S'. Use the hint in the problem. For two vectors in the xy-plane, the cross product has a z component given by $A_xB_y - B_xA_y$, so the magnitude of this quantity is the area of the parellelogram. Apply this approach to the diagram in Fig. 37.67. $\vec{A}$ is the vector from 1 to 3 and $\vec{B}$ is the vector from 1 to 2. Their components are $A_x = -\gamma vt$, $A_y = \gamma T$, $B_x = -\gamma L$, and

$B_y = Lv\gamma/c^2$. The area is $A = |A_xB_y - B_xA_y| = LT$. But recall that we replace t with ct, so $T \to cT$. Therefore the area is $A = cLT$.

EVALUATE: **(e)** The area is cLT in both reference frames. So it is independent of the reference frame, or "Lorentz invariant."

STUDY GUIDE FOR PHOTONS: LIGHT WAVES BEHAVING AS PARTICLES

Summary

We'll explore the particle nature of photons in this chapter. We'll see that the energy of electromagnetic wave is quantized, that it is emitted and absorbed in photons that behave as particle-like packages that carry energy and momentum. We'll examine the discoveries that led to this new interpretation, including the photoelectric effect, Compton scattering, and pair production. We'll also discover that electromagnetic radiation exhibits wave–particle duality.

Objectives

After studying this chapter, you'll understand

- How to determine the energy and momentum of a photon.
- How the photoelectric effect confirmed the photon nature of light.
- How to use the photon nature of light to interpret the Compton scattering of X-rays.
- Wave–particle duality for photons.
- How to apply the Heisenberg uncertainty principle to photons.

Concepts and Equations

Term	Description
Photons	Electromagnetic radiation exhibits both wave and particle behavior. Photons carry units of electromagnetic radiation. The energy of a photon is given by $$E = hf = \frac{hc}{\lambda},$$ where $h = 6.626 \times 10^{-34}$ Js is Planck's constant, f is the frequency of the photon, and λ is the wavelength of the photon. The momentum of a photon is given by $$p = \frac{E}{c} = \frac{hf}{c} = \frac{h}{\lambda}.$$
Photoelectric Effect	The photoelectric effect describes how a photon striking a surface can eject an electron from that surface. The photon must have sufficient energy (greater than the work function ϕ of the material of the surface) for the electron to escape. Mathematically, $$eV_0 = hf - \phi.$$
X-Rays and Photon Scattering	X-rays are high-energy, small-wavelength photons that can be produced when electrons strike a target. X-rays may scatter from electrons bound to a nucleus in a process called Compton scattering. After scattering, the X-rays have less energy and a longer wavelength. The change in an X-ray's wavelength is given by $$\Delta\lambda = \lambda' - \lambda = \frac{h}{mc}(1 - \cos\phi),$$ where m is the electron's mass and ϕ is the scattering angle. This discovery helped prove that light, X-rays, and all electromagnetic radiation are made of discrete energy packets called photons.

Key Concept 1: Photons have both particle properties (energy and momentum) and wave properties (frequency and wavelength). The energy E and the magnitude of momentum p of a photon are both proportional to the frequency and inversely proportional to the wavelength; furthermore, E is directly proportional to p.

Key Concept 2: In an experiment with the photoelectric effect, not all photoelectrons have the same kinetic energy. The *stopping voltage*—the voltage required to stop photoelectrons emitted at the cathode from reaching the anode—tells you the *maximum* kinetic energy of the photoelectrons.

Key Concept 3: To analyze the results of a photoelectric-effect experiment, graph the stopping potential as a function of the frequency of the light used. You can calculate the work function of the material used from the vertical intercept of this graph, and you can calculate the value of Planck's constant from the slope of this graph.

Key Concept 4: In bremsstrahlung, a photon of maximum energy and minimum wavelength λ_{min} is produced when all of the kinetic energy of an electron goes into producing that photon. The value of λ_{min} is inversely proportional to the voltage used to accelerate the electron.

Key Concept 5: In Compton scattering, a photon collides with an electron; both energy and momentum are conserved in the collision. If the electron is initially at rest, the photon loses both energy and momentum to the electron. The scattered photon wavelength is longer than the wavelength of the incident photon by an amount $\Delta\lambda$ that depends on the photon scattering angle.

Key Concept 6: In *pair annihilation,* an electron and a positron collide and disappear, and they are replaced by two or more photons.

Energy and momentum are conserved in pair annihilation.

From Chapter 38 of Student's Study Guide to accompany *University Physics with Modern Physics, Volume 3,* Fifteenth Edition. Hugh D. Young and Roger A. Freedman. Copyright © 2020 by Pearson Education, Inc. All rights reserved.

Key Concept 7: The Heisenberg uncertainty principle takes two forms. First, the smaller the uncertainty in the *location* of a physical system, the greater the uncertainty in the *momentum* of the system. Second, the smaller the uncertainty in the *time* of a physical phenomenon, the greater the uncertainty in the *energy* of the phenomenon.

Conceptual Questions

1: Finding the stopping potential

Figure 1 shows a graph of the stopping potential as a function of the frequency of incident light illuminating a metal surface. Find the photoelectric work function for this surface.

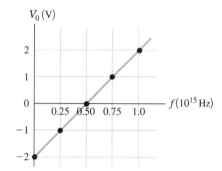

Figure 1 Question 1.

IDENTIFY, SET UP, AND EXECUTE The graph of the stopping potential as a function of the frequency of light is a straight line given by

$$V_0 = \frac{h}{e}f - \frac{\phi}{e}.$$

The vertical intercept of this graph is the negative of the work function, divided by e. Here, the vertical intercept is -2.0 V, so the work function is 2.0 eV.

EVALUATE We'll often graph the stopping potential as a function of the frequency of light in order to find the work function. We see how extracting the work function is simple.

2: Compton effect

Why is there no wavelength shift for forward scattering $(\theta = 0°)$ in the Compton effect?

IDENTIFY, SET UP AND EXECUTE When $\theta = 0°$ is substituted into the Compton scattering equation, the wavelength shift is found to be zero. A photon scattered at $0°$ does not interact with an electron; its momentum and energy remain unchanged. Since the photon does not interact, it continues undeflected with the same wavelength.

EVALUATE The photon's energy and momentum must remain constant due to conservation of energy and momentum.

3: Particles from nowhere

Does the Heisenberg uncertainty principle imply that particles could be created for very short amounts of time? How long could an electron–positron pair exist without violating the uncertainty principle?

IDENTIFY, SET UP, AND EXECUTE The uncertainty principle for energy and time states that the product of the uncertainty in energy and the uncertainty in time is greater than $h/2\pi$. The uncertainty in energy can be very large, as long as the uncertainty in time is very small. The energy needed to create an electron–positron pair is roughly 1 MeV, so the uncertainty in time would be

$$\Delta t \geq \frac{\hbar}{\Delta E} = \frac{(6.626 \times 10^{-34} \text{ J} \cdot \text{s})}{2\pi (10^6 \text{ eV})(1.6 \times 10^{-19} \text{ J/eV})} = 6.6 \times 10^{-22} \text{ s.}$$

If the electron and positron are created and annihilated within 6.6×10^{-22} s, then the uncertainty principle is not violated.

EVALUATE Creating particles from vacuum is not just a theoretical exercise. In what is called the quantum vacuum, particles are created and annihilated on a continual basis. Using particle accelerators, physicists have proven the existence of these particles by scattering other particles off of them during their short lifetimes. These so-called "sea" particles (since they are created in the vacuum "sea") are responsible for important phenomena at the subatomic scale.

Problems

1: Photoelectric effect for sodium

The photoelectric work function for sodium is 2.7 eV. If light of frequency 9.0×10^{14} Hz falls on sodium, find (a) the stopping potential, (b) the kinetic energy of the most energetic electrons ejected, and (c) the speed of those electrons.

IDENTIFY AND SET UP We'll use energy conservation to find the stopping potential. The most energetic electrons will have energy equal to the stopping potential energy.

EXECUTE The stopping potential energy is equal to the photon's energy minus the work function:

$$eV_0 = hf - \phi.$$

The stopping potential is then

$$V_0 = \frac{hf - \phi}{e} = \frac{(9.0 \times 10^{14} \text{ Hz})(6.626 \times 10^{-34} \text{ J} \cdot \text{s}) - (2.7 \text{ eV})(1.6 \times 10^{-19} \text{ J/eV})}{1.6 \times 10^{-19} \text{ C}} = 1.03 \text{ V.}$$

The maximum kinetic energy of an electron is then

$$K = eV_0 = e(1.03 \text{ V}) = 1.03 \text{ eV.}$$

and the speed of that electron is

$$v = \sqrt{\frac{2K}{m}} = \sqrt{\frac{2(1.03 \text{ eV})(1.6 \times 10^{-19} \text{ J/eV})}{(9.11 \times 10^{-31} \text{ kg})}} = 6.01 \times 10^5 \text{ m/s.}$$

The stopping potential is 1.03 V, the maximum kinetic energy of the electrons that are ejected is 1.03 eV, and the speed of the electrons ejected with maximum kinetic energy is $6.01 \times 10^5 \, \text{m/s}$.

KEY CONCEPT 2 **EVALUATE** We rely on basic energy conservation laws to solve problems involving the photoelectric effect. Note how we've used energy relations which imply that light is a particle.

Practice Problem: Will there be electrons ejected when the light falls on aluminum, which has a work function of 4.3 eV? *Answer:* No

Extra Practice: What minimum frequency is required to remove electrons from aluminum? *Answer:* 1.04×10^{15} Hz

2: Stopping potential

The stopping potential for photoelectrons ejected from a surface by 375-nm photons is 1.870 V. Calculate the stopping potential for 600-nm photons.

IDENTIFY AND SET UP We'll use energy conservation to find the work function and then use that to find the stopping potential for the 600-nm photons, the target variable.

EXECUTE The stopping potential energy is equal to the photon's energy minus the work function:

$$eV_0 = hf - \phi.$$

The work function is found from the 375-nm photon response. Substituting and solving gives

$$\phi = \frac{hc}{\lambda} - eV_0$$

$$= \frac{(6.626 \times 10^{-34} \text{ J} \cdot \text{s})(3.00 \times 10^8 \text{ m/s})}{(375 \times 10^{-9} \text{ m})} - (1.6 \times 10^{-19} \text{ C})(1.870 \text{ V})$$

$$= 2.31 \times 10^{-19} \text{ J} = 1.44 \text{ eV}.$$

For the 600-nm photons, we find the stopping potential from

$$eV_0' = \frac{hc}{\lambda'} - \phi$$

$$= \frac{(6.626 \times 10^{-34} \text{ J} \cdot \text{s})(3.00 \times 10^8 \text{ m/s})}{(600 \times 10^{-9} \text{ m})} - (2.31 \times 10^{2\ 19} \text{ J})$$

$$= 1.00 \times 10^{-19} \text{ J}.$$

The stopping potential for the 600-nm photons is then

$$V_0' = \frac{1.00 \times 10^{-19} \text{ J}}{1.6 \times 10^{-19} \text{ C}} = 0.627 \text{ V}.$$

EVALUATE The stopping potential depends on the wavelength of light used, but the work function is constant for a material.

Practice Problem: What is the stopping potential for 375-nm photons when the work function is 2.0 eV? *Answer:* 1.3 V

From Chapter 38 of Student's Study Guide to accompany *University Physics with Modern Physics, Volume 3,* Fifteenth Edition. Hugh D. Young and Roger A. Freedman. Copyright © 2020 by Pearson Education, Inc. All rights reserved.

Extra Practice: What is the stopping potential for 600-nm light in this case? *Answer: 0.07 V*

3: X-ray scattering

X-rays of frequency 9.0×10^{18} Hz Compton-scatter off electrons. What are the wavelength and energy of X-rays scattered to an angle of 135°?

IDENTIFY AND SET UP The Compton-scattering formula gives the change in wavelength for the X-rays under consideration. We'll convert the X-ray frequency to a wavelength before finding the change in wavelength.

EXECUTE The X-rays have wavelength

$$\lambda = \frac{c}{f} = \frac{3.00 \times 10^8 \text{ m/s}}{9.0 \times 10^{18} \text{ Hz}} = 3.33 \times 10^{-11} \text{ m}.$$

The change in wavelength due to Compton scattering is

$$\Delta\lambda = \lambda' - \lambda = \frac{h}{mc}(1 - \cos\phi) = \frac{(6.626 \times 10^{2\,34} \text{ J·s})}{(9.11 \times 10^{-31} \text{ kg})(3.00 \times 10^8 \text{ m/s})}(1 - \cos 135°)$$

$$= 4.15 \times 10^{-12} \text{ m}.$$

The scattered X-ray's wavelength is

$$\lambda' = \lambda + \Delta\lambda = (3.33 \times 10^{-11} \text{ m}) + (4.15 \times 10^{-12} \text{ m}) = 3.75 \times 10^{-11} \text{ m}.$$

The energy of the scattered X-ray is

$$E = \frac{hc}{\lambda} = \frac{(6.626 \times 10^{-34} \text{ J·s})(3.00 \times 10^8 \text{ m/s})}{(3.75 \times 10^{-11} \text{ m})} = 5.31 \times 10^{-15} \text{ J} = 33.2 \text{ keV}.$$

The scattered X-ray has a wavelength of 37.5 pm and an energy of 33.2 keV.

KEY CONCEPT 5 **EVALUATE** Compton scattering changes the wavelength and energy by a relatively small amount. Here, the energy decreased by about 11%. For longer wavelength photons, the effect is much smaller.

Practice Problem: What is the scattered wavelength if the incident light has a frequency of 3×10^{18} Hz instead? *Answer: 104 pm*

Extra Practice: What is the energy of the scattered light? *Answer: 11.9 keV*

4: Compton scattering

Compare the maximum relative frequency change for 550-nm photons to 0.025-nm X-rays. For which type of photon will Compton scattering be more easily observed?

IDENTIFY AND SET UP We'll calculate the maximum frequency shift due to Compton scattering for the two photons. The maximum frequency shift corresponds to a scattering angle of 180°.

EXECUTE The change in wavelength due to Compton scattering is

$$\Delta\lambda = \lambda' - \lambda = \frac{h}{mc}(1 - \cos\phi).$$

Converting the formula to frequencies gives

$$\frac{1}{f'} - \frac{1}{f} = \frac{h}{mc^2}(1 - \cos\phi).$$

Rearranging terms to find the relative frequency change, we obtain

$$\frac{f' - f}{f'} = \frac{\Delta f}{f'} = \frac{hf}{mc^2}(1 - \cos\phi).$$

The maximum frequency shift occurs when the photon is backscattered, or $\phi = 180°$. We find the relative frequency change for the two photons. For the 550-nm photons,

$$\frac{\Delta f}{f'} = \frac{hf}{mc^2}(1 - \cos\phi) = \frac{2h}{mc\lambda} = \frac{2(6.626 \times 10^{-34} \text{ J}\cdot\text{s})}{(9.11 \times 10^{-31} \text{ kg})(3.00 \times 10^8 \text{ m/s})(550 \text{ nm})} = 8.8 \times 10^{-6}.$$

For the 0.025-nm X-rays,

$$\frac{\Delta f}{f'} = \frac{2h}{mc\lambda} = \frac{2(6.626 \times 10^{-34} \text{ J}\cdot\text{s})}{(9.11 \times 10^{-31} \text{ kg})(3.00 \times 10^8 \text{ m/s})(0.025 \text{ nm})} = 0.18.$$

X-ray Compton scattering is much easier to observe than Compton scattering with visible photons.

KEY CONCEPT 5 **EVALUATE** We noted in the previous problem that the Compton effect produces small changes in the scattered photon's energies. We see that effect clearly in the results of this problem, noting that the X-rays have small, but measurable frequency (and energy) changes while the visible-light photons have even smaller, not readily measurable changes in their frequency (and energy).

Try It Yourself!

1: Threshold wavelength

Find the threshold wavelength of light that would produce photoelectrons for a silver surface. The work function for silver is 4.8 eV.

Solution Checkpoints

IDENTIFY AND SET UP Use energy conservation to solve the problem.

EXECUTE Energy conservation for the photoelectric effect gives

$$eV_0 = hf - \phi.$$

At threshold, the photoelectrons have zero kinetic energy. Solving for the wavelength yields

$$\lambda = \frac{hc}{\phi} = 259 \text{ nm}.$$

EVALUATE What is the threshold wavelength for a cesium surface, for which the work function is 1.8 eV?

2: Compton scattering

A 50-keV X-ray strikes an electron at rest and scatters. If the X-ray is scattered at an angle of 90°, find (a) the change in wavelength of the X-ray, (b) the energy of the X-ray after scattering, and (c) the velocity of the electron after scattering.

Solution Checkpoints

IDENTIFY AND SET UP Use the Compton-scattering equation and energy conservation to solve the problem.

EXECUTE (a) Compton scattering gives the change in wavelength for the X-ray:

$$\Delta\lambda = \lambda' - \lambda = \frac{h}{mc}(1-\cos\phi).$$

For this problem, the scattering angle is 90°, the wavelength shift is 2.42×10^{-12} m, and the scattered X-ray has a wavelength of 2.73×10^{-11} m.

(b) The new frequency is found from

$$f' = c/\lambda'.$$

The new energy is 45.6 keV.

(c) The energy lost by the X-ray becomes kinetic energy of the electron. The kinetic energy is much smaller than the electron rest energy, so we can use classical energy expressions:

$$\frac{1}{2}mv^2 = \Delta E = 4.44 \text{ keV}.$$

The electron acquires a velocity of 3.96×10^{7} m/s.

KEY CONCEPT 5 **EVALUATE** Is the electron's velocity small enough to justify using classical energy expressions?

Practice Problem: What is the new energy if 40-keV X-rays are scattered by 90°? *Answer:* 37.1 keV

Extra Practice: What is the new energy if 30-keV X-rays are scattered by 90°? *Answer:* 28.3 keV

Key Example Variation Problems

Solutions to these problems are in Chapter 38 of the Student's Solutions Manual.

Be sure to review EXAMPLES 38.1, 38.2, and 38.3 (Section 38.1) before attempting these problems.

VP38.3.1 A laser emits 2.13×10^{16} photons per second, each of which has wavelength 625 nm. Find (a) the energy and momentum of a single photon and (b) the power output of the laser.

VP38.3.2 A certain metal has a work function of 4.55 eV. For this metal, find (a) the minimum frequency that light must have to produce photoelectrons and (b) the frequency that light must have to produce photoelectrons with maximum kinetic energy 1.53 eV.

VP38.3.3 In a photoelectric-effect experiment, the stopping potential is 1.37 V if the light used to illuminate the cathode has wavelength 475 nm. Find (a) the work function (in eV) of the cathode material and (b) the stopping potential if the wavelength is decreased to 425 nm.

VP38.3.4 Your results from a photoelectric-effect experiment show that the maximum speed of the emitted photoelectrons is 6.95×10^5 m/s when the cathode is illuminated with ultraviolet light of wavelength 306 nm. Find (a) the maximum photoelectron kinetic energy in eV, (b) the photon energy in eV, and (c) the work function.

Be sure to review EXAMPLES 38.5 and 38.6 (Section 38.3) before attempting these problems.

VP38.6.1 You use X-ray photons of wavelength 0.251 nm in a Compton-scattering experiment. At a scattering angle of 14.5°, what is the increase in the wavelength of the scattered X-rays compared to the incident X-rays? Express your results (a) in nm and (b) as a percentage of the wavelength of the incident X-rays.

VP38.6.2 You use X-rays of *frequency* 1.260×10^{18} Hz in a Compton-scattering experiment. Find the frequency of the scattered X-rays if the scattering angle is (a) 90.00° and (b) 180.0°.

VP38.6.3 Compton scattering can also occur if a photon collides with a proton (mass 1.673×10^{-27} kg) at rest. If you do an experiment of this kind using gamma-ray photons with wavelength 2.50×10^{-12} m $= 2.50$ pm, for what scattering angle is the wavelength of the scattered gamma rays longer than the wavelength of the incident gamma rays by (a) 0.0100% and (b) 0.0800%?

VP38.6.4 An antiproton has the same mass and rest energy (938.3 MeV) as a proton, but has a negative change $-e$ instead of the positive charge $+e$ of the proton. A proton and an antiproton, initially far apart, move toward each other with the same speed and collide head-on, annihilating each other and producing two photons. Find the energies and wavelengths of the photons if the initial kinetic energies of the proton and antiproton are (a) both negligible and (b) both 545 MeV.

Be sure to review EXAMPLE 38.7 (Section 38.4) before attempting these problems.

VP38.7.1 A laser produces light of wavelength 633 nm in pulses that propagate in the $+x$-direction and that are 7.00×10^{-6} m in length. For an average photon in this pulse, find (a) the momentum p, (b) the minimum momentum uncertainty in kg·m/s and as a percentage of p, (c) the energy E, and (d) the minimum energy uncertainty in J and as a percentage of E.

VP38.7.2 Weather radar systems emit radio waves in pulses. For a typical system the frequency used is 3.00 GHz (1 GHz $= 10^9$ Hz) and the minimum energy uncertainty of the radio photons is 5.50×10^{-29} J. Find (a) the time duration of each pulse, (b) the length of each pulse, (c) the energy of an average photon, and (d) the minimum frequency uncertainty.

VP38.7.3 In a pulsed laser the minimum energy uncertainty of a photon is 2.33×10^{-20} J, which is 6.70% of the average energy of a photon in the pulse. Find (a) the wave-

From Chapter 38 of Student's Study Guide to accompany *University Physics with Modern Physics, Volume 3,* Fifteenth Edition.
Hugh D. Young and Roger A. Freedman. Copyright © 2020 by Pearson Education, Inc. All rights reserved.

length and frequency of the laser light, (b) the minimum frequency uncertainty, and (c) the time duration of the pulse.

VP38.7.4 The average energy of a photon in a pulsed laser beam is 2.39 eV, with a minimum uncertainty of 0.0155 eV. Each pulse has an average of 5.00×10^{12} photons. Find (a) the time duration of each pulse, (b) the wavelength of the light, and (c) the energy per pulse in J.

VP38.3.1. **IDENTIFY:** This problem involves the energy and momentum of a photon.

SET UP: $E = hc/\lambda,\ \lambda = h/p.$

EXECUTE: **(a)** We want the energy and momentum of the photon. $E = hc/\lambda = hc/(625\ \text{nm}) = 3.18 \times 10^{-19}$ J. $p = h/\lambda = h/(625\ \text{nm}) = 1.06 \times 10^{-27}$ kg·m/s.

(b) We want the power output of the laser. $\dfrac{E}{t} = \dfrac{\left(2.15 \times 10^{16}\, \text{photons}\right)\left(3.18 \times 10^{-19}\, \text{J/photon}\right)}{1.00\ \text{s}}$

$= 6.77$ mW.

EVALUATE: Don't confuse the frequency of the light with the frequency at which photons are emitted. They are *very* different!

VP38.3.2. **IDENTIFY:** This problem is about the photoelectric effect.

SET UP: $hf = K_{\text{max}} + \phi.$

EXECUTE: **(a)** We want the minimum frequency to produced photoelectrons. This is the threshold frequency, at which the kinetic energy is zero, so $hf = \phi.\ f = \phi/h = (4.55\ \text{eV})/h = 1.10 \times 10^{15}$ Hz.

(b) We want the frequency when the maximum kinetic energy is 1.53 eV. Solve $hf = K_{\text{max}} + \phi$ for f, giving

$$f = \frac{K_{\text{max}} + \phi}{h} = \frac{1.53\ \text{eV} + 4.55\ \text{eV}}{h} = 1.47 \times 10^{15}\ \text{Hz}.$$

EVALUATE: Be prepared to use h in units of either J·s or eV·s, depending on the given units. Doing so can avoid tedious unit conversions.

VP38.3.3. **IDENTIFY:** This problem is about the photoelectric effect.

SET UP: $eV_0 = hc/\lambda - \phi,\ f\lambda = c.$

EXECUTE: **(a)** We want the work function. Solve $eV_0 = hc/\lambda - \phi$ for ϕ using $V_0 = 1.37$ V and $\lambda = 475$ nm, giving $\phi = 1.24$ eV.

(b) We want V_0 if $\lambda = 425$ nm. Solve $eV_0 = hc/\lambda - \phi$ for V_0 using $\phi = 1.24$ eV and $\lambda = 425$ nm. This gives $V_0 = 1.68$ eV.

EVALUATE: Decreasing the photon wavelength increases its energy, so the stopping potential increases. From Table 38.1 we see that the ϕ we found is around half that of sodium, so it is a reasonable value.

VP38.3.4. **IDENTIFY:** This problem is about the photoelectric effect.

SET UP: $E = hc/\lambda.$

EXECUTE: **(a)** We want the kinetic energy in eV. Using $K = \frac{1}{2}mv^2$, with m the electron mass and $v = 6.95 \times 10^5$ m/s gives $K = 2.20 \times 10^{-19}$ J $= 1.37$ eV.
(b) We want the photon energy. $E = hc/\lambda = hc/(306 \text{ nm}) = 4.05$ eV.
(c) We want ϕ. $E_{\text{photon}} = K_{\max} + \phi$. $\phi = 4.05$ eV $- 1.37$ eV $= 2.68$ eV.
EVALUATE: The work function is about the same as for sodium, so it is reasonable.

VP38.6.1. **IDENTIFY:** We are dealing with Compton scattering.

SET UP: $\lambda' - \lambda = \frac{h}{mc}(1 - \cos \phi)$. We want the increase in the wavelength.

EXECUTE: **(a)** $\lambda' - \lambda = \frac{h}{mc}(1 - \cos \phi) = \left(2.426 \times 10^{-12} \text{ m}\right)\left(1 - \cos 14.5°\right) = 7.73 \times 10^5$ nm.

(b) $\frac{\lambda' - \lambda}{\lambda} = \frac{7.73 \times 10^5 \text{ nm}}{0.251 \text{ nm}} = 0.0308\%.$

EVALUATE: The fractional change in the wavelength is very small. It would be larger if the scattering angle ϕ were closer to $180°$.

VP38.6.2. **IDENTIFY:** We are dealing with Compton scattering.

SET UP: $\lambda' - \lambda = \frac{h}{mc}(1 - \cos \phi)$, $f\lambda = c$. We want the frequency of the scattered waves. Using $f\lambda = c$ gives

$$\frac{1}{f'} = \frac{1}{f} + \frac{h}{mc^2}(1 - \cos \phi).$$

EXECUTE: **(a)** $\phi = 90.00°$: Using the given frequency and scattering angle gives $f' = 1.247 \times 10^{18}$ Hz.

(b) $\phi = 180.0°$: Working as in (a) but with $\phi = 180.0°$ gives $f' = 1.235 \times 10^{18}$ Hz.
EVALUATE: For $\phi = 180°$ the x rays are scattered directly back from their original direction, but at $\phi = 90.0°$ they go off perpendicular to their original direction. Note that the frequency of the scattered waves is different in each case.

VP38.6.3. **IDENTIFY:** We are dealing with Compton scattering off of a proton.

SET UP: $\lambda' - \lambda = \frac{h}{mc}(1 - \cos \phi)$, where m is now the mass of a *proton*. We want the scattering angle ϕ. Solving for $\cos \phi$ gives $\cos \phi = 1 - \frac{mc}{h}(\lambda' - \lambda)$, with $\lambda = 2.50 \times 10^{-12}$ m.

EXECUTE: **(a)** $\lambda' = \lambda + 0.000100\lambda$: $\cos \phi = 1 - \frac{mc}{h}(\lambda' - \lambda) = \frac{mc}{h}(\lambda + 0.000100\lambda - \lambda)$

$= 1 - \frac{mc}{h}(0.000100\lambda).$

Using the *proton* mass gives $\phi = 35.8°$.

(b) $\lambda' = \lambda + 0.000800\lambda$: $\cos \phi = 1 - \frac{mc}{h}(\lambda' - \lambda) = \frac{mc}{h}(\lambda + 0.000100\lambda - \lambda) = 1 - \frac{mc}{h}(0.000800\lambda).$

$\phi = 121°$.
EVALUATE: Note that the scattering angle has a considerable effect on the wavelength of the scattered wave.

VP38.6.4. **IDENTIFY:** This problem involves the annihilation of a proton and an antiproton.
SET UP: We want the energy and wavelength of the resulting photons. $E = hc/\lambda$.

EXECUTE: (a) Initial kinetic energy is negligible. The energy of the two photons is equal to the rest energy of the two protons (which is 938.3 MeV each), and each photon has the same energy due to momentum conservation. Therefore $E_{photon} = 938.3$ MeV. The wavelength of each photon is $\lambda = hc/E = hc/(938.3$ MeV$) = 1.32 \times 10^{-15}$ m.

(b) Initial kinetic energy of each proton is 545 MeV. In this case, the energy of the photons is equal to the rest energy of the protons plus their kinetic energy, so the energy of each photon is E_{photon} = 938.3 MeV + 545 MeV = 1483.3 MeV. $\lambda = hc/E = hc/(1483.3$ MeV$) = 8.36 \times 10^{-16}$ m.

EVALUATE: The wavelength of the photons is less when the protons have substantial energy. This is a reasonable result because short-wavelength photons have more energy than long-wavelength photons.

VP38.7.1. **IDENTIFY:** This problem involves the uncertainty principle.
SET UP: $\Delta t \Delta E = \hbar/2$, $\Delta x \Delta p_x = \hbar/2$, $\lambda = h/p$, $E = hc/\lambda$.

EXECUTE: (a) We want the momentum. $p = h/\lambda = h/(633$ nm$) = 1.05 \times 10^{-27}$ kg·m/s.

(b) We want the minimum uncertainty in the momentum. We know that $\Delta x = 700 \times 10^{-6}$ m.

$$\Delta p_{min} = \frac{\hbar}{2\Delta x} = \frac{1.055 \times 10^{-34} \text{ J·s}}{2(7.00 \times 10^{-6} \text{ m})} = 7.54 \times 10^{-30} \text{ kg·m/s.}$$

The percent uncertainty is

$$\frac{\Delta p_{min}}{p_{min}} = \frac{7.54 \times 10^{-30} \text{ kg·m/s}}{1.05 \times 10^{-27} \text{ kg·m/s}} = 0.720\%.$$

(c) We want the energy. $E = hc/\lambda = hc/(633$ nm$) = 3.14 \times 10^{-19}$ J.

(d) We want the minimum uncertainty in the energy. We use $\Delta t \Delta E = \hbar/2$ where Δt is the time for the pulse to propagate 7.00×10^{-6} m. Therefore

$$\Delta t = \frac{\Delta x}{c} = \frac{7.00 \times 10^{-6} \text{ m}}{c} = 2.333 \times 10^{-14} \text{ s.}$$

$$\Delta E = \frac{\hbar}{2\Delta t} = \frac{h}{2\left(2.333 \times 10^{-14} \text{ s}\right)} = 2.26 \times 10^{-21} \text{ s.}$$

Dividing the result in (d) by the energy in (c) gives

$$\frac{\Delta E}{E} = 0.720\%.$$

EVALUATE: Note that the uncertainties in p and E are very small percents.

VP38.7.2. **IDENTIFY:** This problem involves the uncertainty principle.
SET UP: $\Delta t \Delta E = h/2$, $E = hf$, $f\lambda = c$.
EXECUTE: (a) We want the time Δt of the pulse.

$$\Delta t = \frac{\hbar}{2\Delta E} = \frac{\hbar}{2\left(5.50 \times 10^{-29} \text{ J}\right)} = 9.59 \times 10^{-7} \text{ s.}$$

(b) We want the length of the pulse. $L = c\Delta t = c\left(9.59 \times 10^{-7} \text{ s}\right) = 288$ m.

(c) We want the energy. $E = hf = h(3.00$ GHz$) = 1.99 \times 10^{-24}$ J.

(d) We want the uncertainty in the frequency. $E = hf$, so $\Delta E = h\Delta f$. Also $\Delta E = \dfrac{h}{2\Delta t}$ $\Delta E = \dfrac{\hbar}{2\Delta t}$.

Combining gives $\dfrac{\hbar}{2\Delta t} = h\Delta f$, so

$$\Delta f = \frac{1}{4\pi\Delta t} = \frac{1}{4\pi\left(9.59\times10^{-7}\text{ s}\right)} = 8.30\times10^{-5}\text{ GHz.}$$

EVALUATE: As we have seen before, the fractional uncertainties arising from the uncertainty principle are usually very small.

VP38.7.3. **IDENTIFY:** This problem involves the uncertainty principle.

SET UP: $\Delta t\Delta E \geq \hbar/2$, $E = hf$, $f\lambda = c$.

EXECUTE: **(a)** We want the wavelength and frequency of the laser light. Using $E = hf$, we see that $\Delta E = h\Delta f$. We know that $\Delta E = 6.70\%E = 0.0670hf$ and that $\Delta E = 2.33 \times 10^{-20}$ J. Therefore 2.33×10^{-20} J $= 0.0670hf$, which gives $f = 5.25 \times 10^{14}$ Hz. Using this frequency, the wavelength is $\lambda = c/f = 572$ nm.

(b) We want the minimum uncertainty in f. $E = hf$, so $\Delta E = h\Delta f$. Thus

$$\Delta f = \frac{\Delta E}{h} = \frac{2.33\times10^{-20}\text{ J}}{h} = 3.52\times10^{13}\text{ Hz.}$$

(c) We want the time duration Δt of the pulse. Using $\Delta t\Delta E \geq \hbar/2$ gives

$$\Delta t = \frac{\hbar}{2\Delta E} = \frac{\hbar}{2(2.33\times10^{-20}\text{ J})} = 2.26\times10^{-15}\text{ s} = 2.26\text{ fs.}$$

EVALUATE: As before, the fractional uncertainties are very small.

VP38.7.4. **IDENTIFY:** This problem involves the uncertainty principle.

SET UP: $\Delta t\Delta E \geq \hbar/2$, $E = hc/\lambda$. The average photon energy is 0.0155 eV, the energy of a pulse is 2.39 eV, and on the *average* there are 5.00×10^{12} photons per pulse.

EXECUTE: **(a)** We want the time duration of a pulse. As pointed out in Example 7.4, the minimum uncertainty Δt for a photon is the time duration of the pulse. Using $\Delta t\Delta E \geq \hbar/2$ gives

$$\Delta t = \frac{\hbar}{2\Delta E} = \frac{\left(4.136\times10^{-15}\text{ eV}\cdot\text{s}\right)/2\pi}{2(0.0155\text{ eV})} = 2.12\times10^{-14}\text{ s.}$$

(b) We want the wavelength. Using $E = hc/\lambda$ gives $\lambda = hc/E = hc/(2.39\text{ eV}) = 519$ nm.

(c) We want the energy per pulse. This energy is equal to the energy per average photon times the number of photons in a pulse. Therefore $E = (2.39\text{ eV/photon})(5.00 \times 10^{12}\text{ photons/pulse})$ $= 1.195 \times 10^{13}$ eV $= 1.91$ μJ per pulse.

EVALUATE: The energy per pulse is small, but the pulse lasts for a very short time, so the power during the pulse is very large. Note that some of the photons have wavelengths longer than 519 and some have shorter wavelengths.

38.3. **IDENTIFY and SET UP:** $c = f\lambda$. The source emits $(0.05)(75\text{ J}) = 3.75$ J of energy as visible light each second. $E = hf$, with $h = 6.63\times10^{-34}$ J·s.

EXECUTE: **(a)** $f = \dfrac{c}{\lambda} = \dfrac{3.00\times10^8\text{ m/s}}{600\times10^{-9}\text{ m}} = 5.00\times10^{14}$ Hz.

(b) $E = hf = (6.63 \times 10^{-34} \text{ J} \cdot \text{s})(5.00 \times 10^{14} \text{ Hz}) = 3.32 \times 10^{-19}$ J. The number of photons emitted per second is $\dfrac{3.75 \text{ J}}{3.32 \times 10^{-19} \text{ J/photon}} = 1.13 \times 10^{19}$ photons.

EVALUATE: **(c)** No. The frequency of the light depends on the energy of each photon. The number of photons emitted per second is proportional to the power output of the source.

38.5. **IDENTIFY** and **SET UP:** A photon has zero rest mass, so its energy is $E = pc$ and its momentum is $p = \dfrac{h}{\lambda}$.

EXECUTE: **(a)** $E = pc = (8.24 \times 10^{-28} \text{ kg} \cdot \text{m/s})(2.998 \times 10^{8} \text{ m/s}) = 2.47 \times 10^{-19}$ J

$= (2.47 \times 10^{-19} \text{ J})(1 \text{ eV}/1.602 \times 10^{-19} \text{ J}) = 1.54$ eV.

(b) $p = \dfrac{h}{\lambda}$, so $\lambda = \dfrac{h}{p} = \dfrac{6.626 \times 10^{-34} \text{ J} \cdot \text{s}}{8.24 \times 10^{-28} \text{ kg} \cdot \text{m/s}} = 8.04 \times 10^{-7}$ m $= 804$ nm.

EVALUATE: This wavelength is longer than visible wavelengths; it is in the infrared region of the electromagnetic spectrum. To check our result we could verify that the same E is given by $E = hc/\lambda$, using the λ we have calculated.

38.7. **IDENTIFY:** The photoelectric effect occurs. The kinetic energy of the photoelectron is the difference between the initial energy of the photon and the work function of the metal.

SET UP: $\frac{1}{2}mv_{max}^2 = hf - \phi$, $E = hc/\lambda$.

EXECUTE: Use the data for the 400.0-nm light to calculate ϕ. Solving for ϕ gives

$\phi = \dfrac{hc}{\lambda} - \frac{1}{2}mv_{max}^2 = \dfrac{(4.136 \times 10^{-15} \text{ eV} \cdot \text{s})(3.00 \times 10^{8} \text{ m/s})}{400.0 \times 10^{-9} \text{ m}} - 1.10 \text{ eV} = 3.10 \text{ eV} - 1.10 \text{ eV} = 2.00$ eV.

Then for 300.0 nm, we have

$\frac{1}{2}mv_{max}^2 = hf - \phi = \dfrac{hc}{\lambda} - \phi = \dfrac{(4.136 \times 10^{-15} \text{ eV} \cdot \text{s})(3.00 \times 10^{8} \text{ m/s})}{300.0 \times 10^{-9} \text{ m}} - 2.00 \text{ eV}$, which gives

$\frac{1}{2}mv_{max}^2 = 4.14 \text{ eV} - 2.00 \text{ eV} = 2.14$ eV.

EVALUATE: When the wavelength decreases the energy of the photons increases and the photoelectrons have a larger minimum kinetic energy.

38.9. **(b)** **IDENTIFY:** Solve part (b) first. First use $eV_0 = hf - \phi$ to find the work function ϕ

SET UP: $eV_0 = hf - \phi$, so $\phi = hf - eV_0 = \dfrac{hc}{\lambda} - eV_0$.

EXECUTE: $\phi = \dfrac{(6.626 \times 10^{-34} \text{ J} \cdot \text{s})(2.998 \times 10^{8} \text{ m/s})}{254 \times 10^{-9} \text{ m}} - (1.602 \times 10^{-19} \text{ C})(0.181 \text{ V})$.

$\phi = 7.821 \times 10^{-19} \text{ J} - 2.900 \times 10^{-20} \text{ J} = 7.531 \times 10^{-19} \text{ J}(1 \text{ eV}/1.602 \times 10^{-19} \text{ J}) = 4.70$ eV.

(a) **IDENTIFY** and **SET UP:** The threshold frequency f_{th} is the smallest frequency that still produces photoelectrons. It corresponds to $K_{max} = 0$ in the equation $\frac{1}{2}mv_{max}^2 = hf - \phi$, so $hf_{th} = \phi$.

EXECUTE: $f = \dfrac{c}{\lambda}$ says $\dfrac{hc}{\lambda_{th}} = \phi$.

$\lambda_{th} = \dfrac{hc}{\phi} = \dfrac{(6.626 \times 10^{-34} \text{ J} \cdot \text{s})(2.998 \times 10^{8} \text{ m/s})}{7.531 \times 10^{-19} \text{ J}} = 2.64 \times 10^{-7}$ m $= 264$ nm.

EVALUATE: As calculated in part (b), $\phi = 4.70$ eV. This is the value given in Table 38.1 for copper.

38.11. IDENTIFY: This problem is about the characteristics of photons.

SET UP: $E = hf$, $\lambda = h/p$, $f\lambda = c$. We want the frequency, wavelength, and momentum of the photon.

EXECUTE: $E_{ph} = 0.700K_{el} = 0.700eV = hf$. So $f = (0.700eV)/h$. Using $V = 50.0$ kV and h in terms of eV, we get $f = 8.46 \times 10^{18}$ Hz. Using this result gives $\lambda = c/f = 0.0355$ nm. Using this wavelength gives
$p = h/\lambda = 1.87 \times 10^{-23}$ kg·m/s.

EVALUATE: The frequency of this photon is much greater than that of visible light.

38.15. IDENTIFY: Apply $\lambda' - \lambda = \dfrac{h}{mc}(1 - \cos\phi) = \lambda_C(1 - \cos\phi)$.

SET UP: Solve for λ': $\lambda' = \lambda + \lambda_C(1 - \cos\phi)$.
The largest λ' corresponds to $\phi = 180°$, so $\cos\phi = -1$.

EXECUTE: $\lambda' = \lambda + 2\lambda_C = 0.0665 \times 10^{-9}$ m $+ 2(2.426 \times 10^{-12}$ m$) = 7.135 \times 10^{-11}$ m $= 0.0714$ nm.
This wavelength occurs at a scattering angle of $\phi = 180°$.

EVALUATE: The incident photon transfers some of its energy and momentum to the electron from which it scatters. Since the photon loses energy its wavelength increases, $\lambda' > \lambda$.

38.17. IDENTIFY and SET UP: The shift in wavelength of the photon is $\lambda' - \lambda = \dfrac{h}{mc}(1 - \cos\phi)$ where λ' is the wavelength after the scattering and $\dfrac{h}{mc} = \lambda_C = 2.426 \times 10^{-12}$ m. The energy of a photon of wavelength λ is $E = \dfrac{hc}{\lambda} = \dfrac{1.24 \times 10^{-6} \text{ eV·m}}{\lambda}$. Conservation of energy applies to the collision, so the energy lost by the photon equals the energy gained by the electron.

EXECUTE:

(a) $\lambda' - \lambda = \lambda_C(1 - \cos\phi) = (2.426 \times 10^{-12}$ m$)(1 - \cos 35.0°) = 4.39 \times 10^{-13}$ m $= 4.39 \times 10^{-4}$ nm.

(b) $\lambda' = \lambda + 4.39 \times 10^{-4}$ nm $= 0.04250$ nm $+ 4.39 \times 10^{-4}$ nm $= 0.04294$ nm.

(c) $\lambda' - \lambda = \left(\dfrac{h}{mc}\right)(1 - \cos\phi) = 0.1050 \times 10^{-9}$ m $+ (2.426 \times 10^{-12}$ m$)(1 - \cos 60.0°) = 0.1062 \times 10^{-9}$ m

and $E_{\lambda'} = \dfrac{hc}{\lambda'} = 2.888 \times 10^4$ eV, so the photon loses 300 eV of energy.

(d) Energy conservation says the electron gains 300 eV of energy.

EVALUATE: The photon transfers energy to the electron. Since the photon loses energy, its wavelength increases.

38.23. IDENTIFY: The uncertainty principle relates the uncertainty in the duration time of the pulse and the uncertainty in its energy, which we know.

SET UP: $E = hc/\lambda$ and $\Delta E \Delta t = \hbar/2$.

EXECUTE: $E = \dfrac{hc}{\lambda} = \dfrac{(6.626 \times 10^{-34} \text{ J·s})(2.998 \times 10^8 \text{ m/s})}{625 \times 10^{-9} \text{ m}} = 3.178 \times 10^{-19}$ J. The uncertainty in the energy is 1.0% of this amount, so $\Delta E = 3.178 \times 10^{-21}$ J. We now use the uncertainty principle.

Solving $\Delta E \Delta t = \dfrac{\hbar}{2}$ for the time interval gives

$$\Delta t = \frac{\hbar}{2\Delta E} = \frac{1.055 \times 10^{-34} \text{ J} \cdot \text{s}}{2(3.178 \times 10^{-21} \text{ J})} = 1.66 \times 10^{-14} \text{ s} = 16.6 \text{ fs}.$$

EVALUATE: The uncertainty in the energy limits the duration of the pulse. The more precisely we know the energy, the longer the duration must be.

38.25. **IDENTIFY** and **SET UP:** The energy added to mass m of the blood to heat it to $T_f = 100°C$ and to vaporize it is $Q = mc(T_f - T_i) + mL_v$, with $c = 4190 \text{ J/kg} \cdot \text{K}$ and $L_v = 2.256 \times 10^6 \text{ J/kg}$. The energy of one photon is $E = \dfrac{hc}{\lambda} = \dfrac{1.99 \times 10^{-25} \text{ J} \cdot \text{m}}{\lambda}$.

EXECUTE: **(a)**

$Q = (2.0 \times 10^{-9} \text{ kg})(4190 \text{ J/kg} \cdot \text{K})(100°C - 33°C) + (2.0 \times 10^{-9} \text{ kg})(2.256 \times 10^6 \text{ J/kg}) =$ 5.07×10^{-3} J. The pulse must deliver 5.07 mJ of energy.

(b) $P = \dfrac{\text{energy}}{t} = \dfrac{5.07 \times 10^{-3} \text{ J}}{450 \times 10^{-6} \text{ s}} = 11.3 \text{ W}.$

(c) One photon has energy $E = \dfrac{hc}{\lambda} = \dfrac{1.99 \times 10^{-25} \text{ J} \cdot \text{m}}{585 \times 10^{-9} \text{ m}} = 3.40 \times 10^{-19}$ J. The number N of photons per pulse is the energy per pulse divided by the energy of one photon:

$$N = \frac{5.07 \times 10^{-3} \text{ J}}{3.40 \times 10^{-19} \text{ J/photon}} = 1.49 \times 10^{16} \text{ photons}.$$

EVALUATE: The power output of the laser is small but it is focused on a small area, so the laser intensity is large.

38.29. **IDENTIFY:** Compton scattering occurs, and we know the angle of scattering and the initial wavelength (and hence momentum) of the incident photon.

SET UP: $\lambda' - \lambda = \left(\dfrac{h}{mc}\right)(1 - \cos\phi)$ and $p = h/\lambda$. Let $+x$ be the direction of propagation of the incident photon and let the scattered photon be moving at 30.0° clockwise from the $+y$-axis.

EXECUTE:

$\lambda' - \lambda = \left(\dfrac{h}{mc}\right)(1 - \cos\phi) = 0.1050 \times 10^{-9} \text{ m} + (2.426 \times 10^{-12} \text{ m})(1 - \cos 60.0°) = 0.1062 \times 10^{-9} \text{ m}.$

$P_{ix} = P_{fx}. \quad \dfrac{h}{\lambda} = \dfrac{h}{\lambda'}\cos 60.0° + p_{ex}.$

$p_{ex} = \dfrac{h}{\lambda} - \dfrac{h}{2\lambda'} = h\dfrac{2\lambda' - \lambda}{(2\lambda')(\lambda)} = (6.626 \times 10^{-34} \text{ J} \cdot \text{s})\dfrac{2.1243 \times 10^{-10} \text{ m} - 1.050 \times 10^{-10} \text{ m}}{(2.1243 \times 10^{-10} \text{ m})(1.050 \times 10^{-10} \text{ m})}.$

$p_{ex} = 3.191 \times 10^{-24} \text{ kg} \cdot \text{m/s}. \quad P_{iy} = P_{fy}. \quad 0 = \dfrac{h}{\lambda'}\sin 60.0° + p_{ey}.$

$p_{ey} = -\dfrac{(6.626 \times 10^{-34} \text{ J} \cdot \text{s})\sin 60.0°}{0.1062 \times 10^{-9} \text{ m}} = -5.403 \times 10^{-24} \text{ kg} \cdot \text{m/s}.$

$p_e = \sqrt{p_{ex}^2 + p_{ey}^2} = 6.28 \times 10^{-24} \text{ kg} \cdot \text{m/s}. \quad \tan\theta = \dfrac{p_{ey}}{p_{ex}} = \dfrac{-5.403}{3.191}$ and $\theta = -59.4°.$

EVALUATE: The incident photon does not give all of its momentum to the electron, since the scattered photon also has momentum.

38.31. IDENTIFY and SET UP: Find the average change in wavelength for one scattering and use that in $\Delta\lambda$ in $\lambda' - \lambda = \left(\dfrac{h}{mc}\right)(1-\cos\phi)$ to calculate the average scattering angle ϕ.

EXECUTE: (a) The wavelength of a 1 MeV photon is

$$\lambda = \frac{hc}{E} = \frac{(4.136\times10^{-15}\text{ eV}\cdot\text{s})(2.998\times10^8\text{ m/s})}{1\times10^6\text{ eV}} = 1\times10^{-12}\text{ m.}$$

The total change in wavelength therefore is $500\times10^{-9}\text{ m} - 1\times10^{-12}\text{ m} = 500\times10^{-9}\text{ m.}$

If this shift is produced in 10^{26} Compton scattering events, the wavelength shift in each scattering event is $\Delta\lambda = \dfrac{500\times10^{-9}\text{ m}}{1\times10^{26}} = 5\times10^{-33}\text{ m.}$

(b) Use this $\Delta\lambda$ in $\Delta\lambda = \dfrac{h}{mc}(1-\cos\phi)$ and solve for ϕ. We anticipate that ϕ will be very small, since $\Delta\lambda$ is much less than h/mc, so we can use $\cos\phi \approx 1 - \phi^2/2.$

$$\Delta\lambda = \frac{h}{mc}\left[1-(1-\phi^2/2)\right] = \frac{h}{2mc}\phi^2.$$

$$\phi = \sqrt{\frac{2\Delta\lambda}{(h/mc)}} = \sqrt{\frac{2(5\times10^{-33}\text{ m})}{2.426\times10^{-12}\text{ m}}} = 6.4\times10^{-11}\text{ rad} = (4\times10^{-9})°.$$

ϕ in radians is much less than 1, so the approximation we used is valid.

(c) IDENTIFY and SET UP: We know the total transit time and the total number of scatterings, so we can calculate the average time between scatterings.

EXECUTE: The total time to travel from the core to the surface is $(10^6\text{ y})(3.156\times10^7\text{ s/y}) = 3.2\times10^{13}\text{ s.}$ There are 10^{26} scatterings during this time, so the average time between scatterings is $t = \dfrac{3.2\times10^{13}\text{ s}}{10^{26}} = 3.2\times10^{-13}\text{ s.}$

The distance light travels in this time is $d = ct = (3.0\times10^8\text{ m/s})(3.2\times10^{-13}\text{ s}) = 0.1\text{ mm.}$

EVALUATE: The photons are on the average scattered through a very small angle in each scattering event. The average distance a photon travels between scatterings is very small.

38.35. IDENTIFY and SET UP: Apply the photoelectric effect. $eV_0 = hf - \phi$. For a photon, $f\lambda = c$.

EXECUTE: (a) Using $eV_0 = hf - \phi$ and $f\lambda = c$, we get $eV_0 = hc/\lambda - \phi$. Solving for V_0 gives $V_0 = \dfrac{hc}{e}\cdot\dfrac{1}{\lambda} - \dfrac{\phi}{e}$. Therefore a graph of V_0 versus $1/\lambda$ should be a straight line with slope equal to hc/e and y-intercept equal to $-\phi/e$. Figure 38.35 shows this graph for the data given in the problem.

The best-fit equation for this graph is $V_0 = (1230\text{ V}\cdot\text{nm})\cdot\dfrac{1}{\lambda} - 4.76\text{ V.}$ The slope is equal to $1230\text{ V}\cdot\text{nm}$, which is equal to $1.23\times10^{-6}\text{ V}\cdot\text{m,}$ and the y-intercept is -4.76 V.

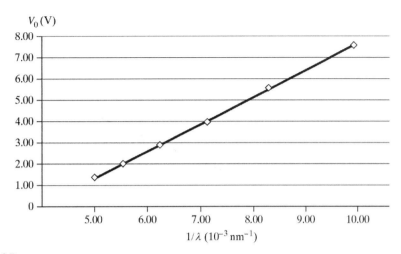

Figure 38.35

(b) Using the slope we have hc/e = slope, so
$h = e(\text{slope})/c = (1.602 \times 10^{-19} \text{ C})(1.23 \times 10^{-6} \text{ V} \cdot \text{m})/(2.998 \times 10^8 \text{ m/s}) = 6.58 \times 10^{-34} \text{ J} \cdot \text{s}$.
The y-intercept is equal to $-\phi/e$, so
$$\phi = -e(y\text{-intercept}) = -(1.602 \times 10^{-19} \text{ C})(-4.76 \text{ V}) = 7.63 \times 10^{-19} \text{ J} = 4.76 \text{ eV}.$$
(c) For the longest wavelength light, the energy of a photon is equal to the work function of the metal, so $hc/\lambda = \phi$. Solving for λ gives $\lambda = hc/\phi$. Our calculation of h was just a test of the data, so we use the accepted value for h in the calculation.
$\lambda = hc/\phi = (6.626 \times 10^{-34} \text{ J} \cdot \text{s})(2.998 \times 10^8 \text{ m/s})/(7.63 \times 10^{-19} \text{ J}) = 2.60 \times 10^{-7} \text{ m} = 260 \text{ nm}$.
(d) The energy of the photon is equal to the sum of the kinetic energy of the photoelectron and the work function, so $hc/\lambda = K + \phi$. This gives $(4.136 \times 10^{-15} \text{ eV} \cdot \text{s})(2.998 \times 10^8 \text{ m/s})/\lambda = 10.0 \text{ eV} + 4.76 \text{ eV} = 14.76 \text{ eV}$, which gives $\lambda = 8.40 \times 10^{-8} \text{ m} = 84.0 \text{ nm}$.
EVALUATE: As we know from Table 38.1, typical metal work functions are several eV, so our results are plausible.

38.37. **IDENTIFY** and **SET UP:** We have Compton scattering, so $\lambda' - \lambda = \left(\dfrac{h}{mc}\right)(1 - \cos\phi)$, which can also be expressed as $\lambda' - \lambda = \lambda_{\text{C}}(1 - \cos\phi)$, where λ_{C} is the Compton wavelength.

EXECUTE: **(a)** Figure 38.37 shows the graph of λ' versus $1 - \cos\phi$ for the data included in the problem. The best-fit equation of the line is $\lambda' = 5.21 \text{ pm} + (2.40 \text{ pm})(1 - \cos\phi)$. The slope is 2.40 pm and the y-intercept is 5.21 pm.

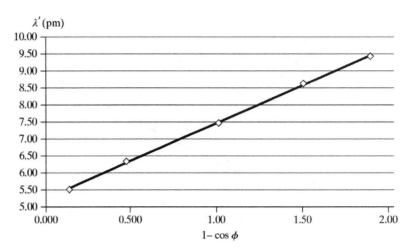

Figure 38.37

(b) Solving $\lambda' - \lambda = \lambda_C (1 - \cos\phi)$ for λ' gives $\lambda' = \lambda + \lambda_C (1 - \cos\phi)$. The graph of λ' versus $1 - \cos\phi$ should be a straight line with slope equal to λ_C and y-intercept equal to λ. From the slope, we get λ_C = slope = 2.40 pm.

(c) From the y-intercept we get $\lambda = y$-intercept = 5.21 pm.

EVALUATE: For backscatter, the photon wavelength would be 5.21 pm + 2(2.40 pm) = 10.01 pm.

STUDY GUIDE FOR PARTICLES BEHAVING AS WAVES

Summary

We'll explore the wave nature of particles in this chapter. Wave–particle duality has shown us that light behaves both like a wave and like a particle. We'll investigate how subatomic particles behave like waves in this chapter, opening our exploration of quantum mechanics. This finding in turn will lead us to the discovery that electrons can behave as waves. In essence, we'll be learning about the very unusual and nonintuitive atomic world. We'll examine the discoveries that led to this new interpretation, including the atomic spectra, the Bohr model of the atom, and blackbody radiation.

Objectives

After studying this chapter, you'll understand

- How electrons and other subatomic particles behave like waves.
- How atomic line spectra reveal the energy levels of atoms.
- The Bohr model of the hydrogen atom and how energy levels are quantized.
- How the laser operates.
- How blackbody radiation shows that electromagnetic radiation is quantized.
- How to use the Heisenberg uncertainty principle to interpret atomic phenomena.

Concepts and Equations

Term	Description
De Broglie Waves	Electrons and other particles have wave properties. Particles are described as waves having a de Broglie wavelength $$\lambda = \frac{h}{p} = \frac{h}{mv}.$$ As waves, particles are inherently spread-out entities described by their wave functions.
Electron Diffraction	The diffraction of an electron beam from the surface of a metallic crystal confirms the wave nature of particles. The wavelength of a nonrelativistic electron accelerated by a potential difference V is given by $$\lambda = \frac{h}{p} = \frac{h}{\sqrt{2meV}}.$$ Images from electron microscopes have much higher resolution than images from visible light, due to the small wavelength of the electrons.
Energy levels and Atomic Line Spectra	An atom making a transition from a higher energy E_i to a lower energy E_f will emit the energy difference through a photon energy: $$hf = \frac{hc}{\lambda} = E_i - E_f.$$ Energy differences can be detected through an atom's spectral lines. For hydrogen, the energy levels are given by $$E_n = -\frac{hcR}{n^2} = -\frac{13.6\ eV}{n^2}, \quad n = 1, 2, 3, \ldots,$$ where $R = 1.097 \times 10^7$ /m is the Rydberg constant.
The Nuclear Atom and the Bohr Model	Rutherford discovered that an atom has a very small positively charged nucleus at its center. Bohr successfully modeled the hydrogen atom as having a lone proton as its nucleus, surrounded by an electron that revolved in certain allowed (quantized) orbits. The electron's allowed angular momenta are given by $$L = mv_n r_n = n\frac{h}{2\pi}, \quad n = 1, 2, 3, \ldots,$$ when n is the principal quantum number. The electron's radius and orbital speed are also quantized: $$r_n = \varepsilon_0 \frac{n^2 h^2}{\pi m e^2} = n^2 a_0 = n^2 (5.29 \times 10^{-11}\ \text{m}),$$ $$v_n = \frac{e^2}{\varepsilon_0 2nh} = \frac{21.9 \times 10^6\ \text{m/s}}{n}.$$
The Laser	The laser operates on the principle of stimulated emission, in which many photons with identical wavelengths and phases are emitted. For the laser to operate, a nonequilibrium population inversion must exist in which more atoms are in a higher-energy state than a lower-energy state. The stimulated emission occurs as the higher-energy state decays to the lower-energy state, triggered by photons passing through the material.

From Chapter 39 of Student's Study Guide to accompany *University Physics with Modern Physics, Volume 3*, Fifteenth Edition.
Hugh D. Young and Roger A. Freedman. Copyright © 2020 by Pearson Education, Inc. All rights reserved.

Blackbody Radiation	The total radiated intensity from a blackbody surface is described by the Stefan-Boltzmann law $$I = \sigma T^4,$$ where $\sigma = 5.67 \times 10^{-8}$ W/m^2 is the Stefan-Boltzmann constant.
Heisenberg Uncertainty Principle	Heisenberg's uncertainty principle states that one cannot determine both the precise position and the precise momentum (or the precise energy and the precise time) of a particle. The uncertainties in each of the two quantities are related as $$\Delta x \Delta p_x \geq \hbar, \qquad \Delta E \Delta t \geq \hbar, \qquad \hbar = \frac{h}{2\pi}.$$ Similar expressions hold for the y- and z-components.

Key Concept 1: Electrons and other subatomic particles have both particle properties (kinetic energy and momentum) and wave properties (frequency and wavelength). Like a photon, the wavelength of a particle is inversely proportional to its momentum; unlike a photon, the wavelength of a nonrelativistic particle is inversely proportional to the square root of its kinetic energy.

Key Concept 2: For a nonrelativistic particle, the speed is proportional to the momentum and hence inversely proportional to the particle's wavelength.

Key Concept 3: The kinetic energy K of an electron equals the magnitude e of its charge multiplied by the voltage V_{ba} required to accelerate it from rest. If the electron is nonrelativistic, both K and V_{ba} are inversely proportional to the square of the particle's wavelength.

Key Concept 4: In Rutherford scattering, a positively charged alpha particle is scattered by a positively charged atomic nucleus. You can find the distance of closest approach using conservation of total mechanical energy (the kinetic energy plus the electric potential energy of the interaction between the alpha particle and the nucleus).

Key Concept 5: When an atom makes a transition between two energy levels, it either emits a photon (if the atom loses energy) or absorbs a photon (if the atom gains energy). The energy of the photon that is emitted or absorbed equals the difference in energy between the two levels. The greater the photon energy, the shorter its wavelength.

Key Concept 6: In the Bohr model of the hydrogen atom, the electron moves around the nucleus in a circular Newtonian orbit. However, only certain orbital radii are allowed. The nth allowed orbit has a distinct kinetic energy K_n, electric potential energy U_n, and total mechanical energy $E_n = K_n + U_n$. When the electron makes a transition between two allowed orbits, a photon is emitted or absorbed with an energy equal to the difference between the values of E_n for the two orbits.

Key Concept 7: The continuous spectrum of light emitted by a blackbody is maximum at a wavelength λ_{max} given by the Wien displacement law: The greater the absolute temperature T of the blackbody, the smaller the value of λ_{max}. The total power radiated from a unit surface area of the blackbody is proportional to T^4.

Key Concept 8: The Planck radiation law gives the spectral emittance (power per unit area per unit wavelength interval) $I(\lambda)$ for an ideal blackbody. To find the power per unit area emitted over a range of wavelengths, integrate $I(\lambda)$ over that range.

Key Concept 9: The Heisenberg uncertainty principle for position and momentum applies to particles as well as to photons. The smaller the uncertainty in the position of a particle, the greater the uncertainty in the particle's momentum. Consequently, a particle confined to a small region of space could have a very large magnitude of momentum and a very large kinetic energy.

Key Concept 10: The Heisenberg uncertainty principle for energy and time interval states that the shorter the duration of an excited state of a system, the greater the minimum uncertainty in the energy of that state.

From Chapter 39 of Student's Study Guide to accompany *University Physics with Modern Physics, Volume 3,* Fifteenth Edition.
Hugh D. Young and Roger A. Freedman. Copyright © 2020 by Pearson Education, Inc. All rights reserved.

Conceptual Questions

1: Interpreting an energy-level diagram

The energy-level diagram of the hypothetical one-electron element nerdium is shown in Figure 1. The potential energy is taken to be zero for an electron an infinite distance from the atom. If a 10.5-eV photon interacts with the nerdium atom in its ground state, what will happen? If a 6.5-eV photon interacts with the nerdium atom in its ground state, what will happen?

$$n = 4 \text{———————} -2.3 \text{ eV}$$
$$n = 3 \text{———————} -4.5 \text{ eV}$$

$$n = 2 \text{———————} -7.5 \text{ eV}$$

$$n = 1 \text{———————} -15.0 \text{ eV}$$

Figure 1 Question 2.

IDENTIFY, SET UP, AND EXECUTE The 10.5-eV photon has energy equal to the energy difference between the $n=1$ and the $n=3$ levels of nerdium. This photon will be absorbed, leaving the nerdium in the $n=3$ excited state.

The 6.5-eV photon has energy less than the energy difference between the $n=1$ and the $n=2$ levels of nerdium. This photon will not be absorbed by the nerdium atom.

EVALUATE We see that atoms do not behave classically. In a classical system, both photons would be absorbed. In the quantized system, only photons with the proper energy difference are absorbed.

Problems

1: Wavelength of a golf ball

An uncharged golf ball of mass 0.1 kg is put into orbit at the earth's surface. Find the de Broglie wavelength of the golf ball.

IDENTIFY AND SET UP The de Broglie wavelength is related to the momentum. We'll find the momentum and then the wavelength.

EXECUTE The force acting on the golf ball in orbit is the gravitational force between the ball and the earth. The ball undergoes centripetal acceleration, so

$$\frac{GMm}{r^2} = mg = \frac{mv^2}{r}.$$

We use this expression to find the velocity of the golf ball. Rearranging terms to solve for the velocity gives

$$v = \sqrt{gR_E},$$

where R_E is the radius of the earth. The momentum is the mass of the golf ball times the velocity. The wavelength is then

From Chapter 39 of Student's Study Guide to accompany *University Physics with Modern Physics, Volume 3,* Fifteenth Edition.
Hugh D. Young and Roger A. Freedman. Copyright © 2020 by Pearson Education, Inc. All rights reserved.

$$\lambda = \frac{h}{p} = \frac{h}{mv} = \frac{h}{m\sqrt{gR_E}}$$

$$= \frac{6.626 \times 10^{-34} \text{ Js}}{(0.1 \text{ kg})\sqrt{(9.8 \text{ m/s}^2)(6.4 \times 10^6 \text{ m})}} = 8.37 \times 10^{-37} \text{ m}.$$

KEY CONCEPT 1 **EVALUATE** We see that the wavelength is much smaller than the size of an atomic nucleus (around 10^{-15} m), so you don't have to worry about the wave nature of a golf ball the next time you play a round of miniature golf!

Practice Problem: What is the de Broglie wavelength of a bowling ball (mass of 7.3 kg) in orbit? *Answer:* 1.15×10^{-38} m

Extra Practice: What is the de Broglie wavelength of a partially loaded Boeing 747 (mass of 300,000 kg) in orbit? *Answer:* 2.8×10^{-43} m

2: Quantum number of a golf ball

Suppose the uncharged golf ball (of mass 0.1 kg) of Problem 1 is viewed as a quantum object in orbit around a nucleus. If its angular momentum is quantized as in the Bohr atom, what is the associated quantum number?

IDENTIFY AND SET UP We'll apply the Bohr model to the golf ball orbiting the earth to solve the problem.

EXECUTE Quantizing the angular momentum gives
$$mvr = n\hbar.$$

In the previous problem, we found that
$$v = \sqrt{gR_E}.$$

Combining these results, we find the following expression for the quantized radius:
$$r_n = \frac{n^2\hbar^2}{m^2 R_E^2 g}.$$

For $r_n = R_E$, we find the value of *n:*

$$n = \sqrt{\frac{m^2 R_E^3 g}{\hbar^2}}$$

$$= \sqrt{\frac{(0.1 \text{ kg})^2 (6.4 \times 10^6 \text{ m})^3 (9.8 \text{ m/s}^2)}{(1.05 \times 10^{-34} \text{ Js})^2}} = 4.8 \times 10^{43}.$$

EVALUATE With this large value of *n*, we see how the allowed energy levels and values of *r* are essentially continuous. It should be clear at this point why the quantum nature of the universe was not discovered in the macroscopic world.

3: A thermal neutron

A thermal neutron is a neutron with mean kinetic energy of $3/2k_BT$, where k_B is Boltzmann's constant and *T* is room temperature (300 K). What is the wavelength of a thermal neutron?

From Chapter 39 of Student's Study Guide to accompany *University Physics with Modern Physics, Volume 3,* Fifteenth Edition. Hugh D. Young and Roger A. Freedman. Copyright © 2020 by Pearson Education, Inc. All rights reserved.

IDENTIFY AND SET UP We'll find the thermal neutron's momentum and then determine the de Broglie wavelength of the neutron.

EXECUTE The kinetic energy of the neutron is given by

$$K = \frac{p^2}{2m} = \frac{3}{2}k_B T.$$

Substituting for the momentum, we have

$$\frac{1}{2m}\left(\frac{h}{\lambda}\right)^2 = \frac{3}{2}k_B T.$$

Solving for λ gives

$$\lambda = \frac{h}{\sqrt{3mk_B T}}$$

$$= \frac{(6.626 \times 10^{-34} \text{ Js})}{\sqrt{3(1.67 \times 10^{-27} \text{ kg})(1.38 \times 10^{-23} \text{ J/K})(300 \text{ K})}} = 0.145 \text{ nm.}$$

KEY CONCEPT 1 **EVALUATE** Thermal neutrons have very small wavelengths. Consequently, they are used in research laboratories to explore the atomic structure of materials through diffraction.

4: Transitions in the hydrogen atom

A hydrogen atom initially in the ground state absorbs a photon, exciting it to the $n = 5$ state. Later, the atom makes a transition to the $n = 2$ state. Find the wavelength of the photon that is absorbed when the atom goes from the ground state to the $n = 5$ state. Then find the wavelength of the photon that is emitted when the atom goes from the $n = 5$ state to the $n = 2$ state.

IDENTIFY We'll use the energy difference to find the wavelength of the photon.

SET UP The photon that is absorbed must have energy equal to the energy difference between the ground state ($n = 1$) and the $n = 5$ state. The transition back to the $n = 2$ state will decrease the energy, so a photon is emitted. This photon's energy will be equal to the energy difference between the $n = 5$ and $n = 2$ states. We'll use that energy to find the wavelength of the emitted photon.

EXECUTE The energy difference between the ground state ($n = 1$) and the $n = 5$ state is

$$E_i - E_f = -\frac{13.6 \text{ eV}}{n^2} - \left(-\frac{13.6 \text{ eV}}{n^2}\right) = -\frac{13.6 \text{ eV}}{1^2} + \frac{13.6 \text{ eV}}{5^2} = -13.1 \text{ eV} = -2.09 \times 10^{-18} \text{ J.}$$

The negative sign indicates that energy is absorbed. The photon's wavelength is

$$\lambda = \frac{hc}{E_i - E_f} = \frac{(6.626 \times 10^{2\,34} \text{ J} \cdot \text{s})(3.00 \times 10^8 \text{ m/s})}{2.09 \times 10^{-18} \text{ J}} = 95 \text{ nm.}$$

The energy difference in the transition from the $n = 5$ state to the $n = 2$ state is

$$E_i - E_f = -\frac{13.6 \text{ eV}}{n^2} - \left(-\frac{13.6 \text{ eV}}{n^2}\right) = -\frac{13.6 \text{ eV}}{5^2} + \frac{13.6 \text{ eV}}{2^2} = 2.86 \text{ eV} = 4.57 \times 10^{-19} \text{ J.}$$

The positive sign indicates that energy is emitted. This photon's wavelength is

$$\lambda = \frac{hc}{E_i - E_f} = \frac{(6.626 \times 10^{-34} \text{ J} \cdot \text{s})(3.00 \times 10^8 \text{ m/s})}{4.57 \times 10^{-19} \text{ J}} = 435 \text{ nm}.$$

A 95-nm-wavelength photon is absorbed to excite the ground-state hydrogen atom to the $n = 5$ state. A 435-nm-wavelength photon is emitted to "de-excite" the excited-state hydrogen atom to the $n = 2$ state.

KEY CONCEPT 5 **EVALUATE** Imagine that instead of predicting the wavelengths of the photons, you were given the wavelengths and had to deduce the energies. This is what physicists did in the early 1900s to unravel the structure of the atom.

Practice Problem: What is the wavelength of the photon that is emitted as the electron returns to the ground state? *Answer:* 121.6 nm

Extra Practice: What wavelength is absorbed when the electron transitions from the ground state to the $n = 3$ state? *Answer:* 102.6 nm

5: Lifetime of a molecule

The full-width, half-maximum intensity for a spectral line characteristic of a pH^2 molecule in an excited rotational energy level is 6×10^9 Hz. Estimate the lifetime of the molecule in this state.

IDENTIFY AND SET UP We'll use the Heisenberg uncertainty principle to solve the problem.

EXECUTE The uncertainty relationship is given by

$$(\Delta E)(\Delta t) \geq \hbar.$$

With this expression, we can estimate the lifetime of the state if we know the energy associated with the state. The energy is found from the formula

$$\Delta E = h \Delta f.$$

Rearranging terms and solving for the lifetime gives

$$\Delta t \geq \frac{\hbar}{h \Delta f} = \frac{1}{2\pi \Delta f} = \frac{1}{2\pi(6 \times 10^9 \text{ /s})} = 2.65 \times 10^{-11} \text{ s}.$$

KEY CONCEPT 10 **EVALUATE** This problem illustrates how we can use the uncertainty principle to estimate the lifetimes of atomic states.

Try It Yourself!

1: Electron diffraction

An electron is accelerated through a potential difference of 1000 V and passes through a thin slit before striking a photographic film 0.5 m away. What should the size of the slit be in order for the first minimum in the electron diffraction pattern to be 0.1 mm from the center of the pattern?

Solution Checkpoints

IDENTIFY AND SET UP Use diffraction to solve the problem. How do you calculate the wavelength of the electron?

EXECUTE The wavelength of the electron is

$$\lambda = \frac{h}{\sqrt{2meV}} = 3.88 \times 10^{-11} \text{ m.}$$

The first minimum is found when

$$\sin\theta = \frac{\lambda}{d}.$$

Using the small-angle approximation, we find that the width is 194 nm.

KEY CONCEPT 3 **EVALUATE** How could a slit of width 194 nm be created?

Practice Problem: What does the size of the slit need to be if the potential difference is 2000 V? *Answer:* 137 nm

Extra Practice: Where would the second minimum be located in this case? *Answer:* 0.2 mm

2: Electron orbiting the nucleus

Calculate (a) the speed of an electron whose wavelength is $2\pi r_0$ (the circumference of its orbit, where $r_0 = 0.053$ nm) and (b) the speed of an electron in the first Bohr orbit of a hydrogen atom.

Solution Checkpoints

IDENTIFY AND SET UP Use the de Broglie wavelength to find the speed for part (a). Use the velocity from the Bohr model to solve part (b).

EXECUTE (a) The de Broglie wavelength leads to

$$v = \frac{h}{m\lambda} = \frac{h}{m2\pi r_0}.$$

Substituting values, we find that the velocity is 2.19×10^6 m/s.
  (b) According to the Bohr model, the velocity is

$$v_n = \frac{e^2}{2\epsilon_0 hn}.$$

Substituting values for $n = 1$, we again find that the velocity is 2.19×10^6 m/s.

EVALUATE Why are the two values the same?

Practice Problem: What is the velocity of an electron in the $n = 2$ state? *Answer:* 1.09×10^6 m/s

Extra Practice: What is the radius of the electron's orbit when its speed is 3.64×10^5 m/s ?
Answer: 1.9×10^{-9} m

3: Atomic spectra

An atom has, in addition to the ground-state energy E_0 at zero energy, energy levels at $E_1 = 10.20$ eV, $E_2 = 12.09$ eV, and $E_3 = 12.75$ eV. If the atom is excited from the ground state to the state with an energy of 12.75 eV, find all possible wavelengths in the atom's spectrum.

Solution Checkpoints

IDENTIFY AND SET UP Find all possible transitions between the various levels, and calculate the corresponding wavelengths.

EXECUTE For any level, the energy is given by

$$\Delta E = \frac{hc}{\lambda}.$$

The excited atom in the E_3 level decays to the E_2, E_1, and E_0 levels, by emitting photons with wavelengths of 1880 nm, 487 nm, and 97.3 nm, respectively. The E_2 level decays to the E_1 and E_0 levels, emitting photons with wavelengths of 657 nm and 102 nm, respectively. Finally, the E_1 level decays to the E_0 level, emitting a photon with wavelength of 122 nm.

KEY CONCEPT 5 **EVALUATE** To calculate the complete spectrum, you must consider all possible decays and their corresponding wavelengths. You may also consider the opposite situation, in which you are given the spectrum and need to calculate the possible levels, much as the physicists who first disentangled atomic spectra had to do.

Key Example Variation Problems

Solutions to these problems are in Chapter 39 of the Student's Solutions Manual.

Be sure to review EXAMPLES 39.1 and 39.2 (Section 39.1) before attempting these problems.

VP39.2.1 In an electron-diffraction experiment, the spacing between atoms in the target is 0.172 nm and the electrons have negligible kinetic energy before being accelerated by an accelerating voltage of 69.0 V. Find (a) the de Broglie wavelength of each electron and (b) the smallest scattering angle for which there is a maximum in the diffraction pattern.

VP39.2.2 The $m = 1$ intensity maximum in an electron-diffraction experiment occurs at an angle of 48.0°. The accelerating voltage of the electrons is 36.5 V. Find (a) the kinetic energy of each electron in joules, (b) the electron de Broglie wavelength, and (c) the spacing between atoms in the target.

VP39.2.3 Using a target with a spacing between atoms of 0.218 nm, you want to produce an electron-diffraction pattern for which the $m = 2$ intensity maximum occurs at an angle of 75.0°. Find (a) the required electron de Broglie wavelength and (b) the required accelerating voltage.

VP39.2.4 A proton (charge e, mass 1.67×10^{-27} kg) is accelerated from rest to 2.38×10^5 m/s. Find (a) the proton de Broglie wavelength and (b) the accelerating voltage.

Be sure to review EXAMPLES 39.5 and 39.6 (Section 39.3) before attempting these problems.

VP39.6.1   A hypothetical atom has energy levels at 0.00 eV, 2.00 eV, and 5.00 eV. If the atom starts in the 5.00 eV level, find all the wavelengths that the atom could possibly emit in the process of returning to the ground level.

VP39.6.2   A hypothetical atom emits a photon of 674 nm when it transitions from the second excited level to the first excited level, and a photon of 385 nm when it transitions from the first excited level to the ground level. Find (a) the energies of the first and second excited levels relative to the ground level and (b) the wavelength of the photon emitted when the atom transitions from the second excited energy level to the ground level.

VP39.6.3   Calculate the energy and the wavelength of the photon emitted by a hydrogen atom when it makes a transition from (a) the $n = 5$ level to the $n = 3$ level, (b) the $n = 4$ level to the $n = 2$ level, and (c) the $n = 3$ level to the $n = 1$ level.

VP39.6.4   A hydrogen atom makes a transition from the $n = 2$ level to the $n = 6$ level. In the Bohr model, find (a) the change in electron kinetic energy, (b) the change in electric potential energy, and (c) the wavelength of the photon that the atom absorbs to cause the transition.

Be sure to review EXAMPLES 39.7 and 39.8 (Section 39.5) before attempting these problems.

VP39.8.1   The star Betelgeuse has surface temperature 3590 K and can be regarded as a blackbody. (a) Find the wavelength at which Betelgeuse emits most strongly. Is this visible, ultraviolet, or infrared? (b) Find the amount of power radiated per unit area of the surface of Betelgeuse.

VP39.8.2   The total power radiated per unit area from a blackbody is 78.0 MW/m^2. Find (a) the temperature of the blackbody and (b) the wavelength at which the blackbody emits most strongly. (c) Is the wavelength in part (b) visible, ultraviolet, or infrared?

VP39.8.3   The spectral emittance curve for the star Rigel A is a good approximation of the curve for a blackbody. The maximum of this curve is at 239 nm. Find (a) the surface temperature of Rigel A and (b) the power that Rigel A radiates per unit surface area.

VP39.8.4   Proxima Centauri, the nearest star to our sun, has a surface temperature of 3040 K. Find (a) the wavelength at which this star emits most strongly and (b) the power that this star radiates per unit surface area in a range of wavelengths 12.0 nm wide centered on the wavelength in part (a).

STUDENT'S SOLUTIONS MANUAL FOR PARTICLES BEHAVING AS
WAVES

VP39.2.1. **IDENTIFY:** This problem involves electron diffraction and the de Broglie wavelength.

SET UP: $\lambda = h/p$, $K = p^2/2m$, $a\sin\theta = m\lambda$, $K = eV$.

EXECUTE: **(a)** We want the de Broglie wavelength. First use $K = eV$ and $K = p^2/2m$ to find p, and then use $\lambda = h/p$ to find λ.

$$p = \sqrt{2mK} = \sqrt{2meV} \,.\, \lambda = \frac{h}{p} = \frac{h}{\sqrt{2meV}}.$$

Using $V = 69.0$ V gives $\lambda = 0.148$ nm.

(b) We want the minimum angle at which a diffraction maximum occurs. Using $m = 1$, $a\sin\theta = m\lambda$ gives $\theta_{min} = \arcsin(\lambda/d) = \arcsin(0.148/0.172) = 59.1°$.

EVALUATE: There is no other angle at which a maximum occurs. We can use $K = p^2/2m$ because the electron is nonrelativistic at this energy which is much less than its rest energy of 0.511 MeV.

VP39.2.2. **IDENTIFY:** This problem involves electron diffraction and the de Broglie wavelength.

SET UP: $\lambda = h/p$, $K = p^2/2m$, $d\sin\theta = m\lambda$, $K = eV$.

EXECUTE: **(a)** We want the kinetic energy. $K = eV = e(36.5 \text{ V}) = 36.5 \text{ eV} = 5.85 \times 10^{-18}$ J.

(b) We want the de Broglie wavelength.

$$\lambda = \frac{h}{p} = \frac{h}{\sqrt{2mK}} = 0.202 \text{ nm}.$$

(c) We want the atomic spacing d. $d\sin\theta = m\lambda$ gives $d = (0.203 \text{ nm})/(\sin 48.0°) = 0.273$ nm.

EVALUATE: We can use $K = p^2/2m$ because the electron's speed is much less than c.

VP39.2.3. **IDENTIFY:** This problem involves electron diffraction and the de Broglie wavelength.

SET UP: $\lambda = h/p$, $K = p^2/2m$, $d\sin\theta = m\lambda$, $K = eV$.

EXECUTE: **(a)** We want the de Broglie wavelength. Solving $d\sin\theta = m\lambda$ with $m = 2$ gives
$\lambda = (0.218 \text{ nm})(\sin 75.0°)/2 = 0.105$ nm.

(b) We want the accelerating voltage V. Solving $eV = K = p^2/2m$ and using $\lambda = h/p$ gives

$$V = \frac{(h/\lambda)^2}{2me} = \frac{h^2}{2me\lambda^2} = \frac{h^2}{2me(0.105 \text{ nm})^2} = 136 \text{ V}.$$

EVALUATE: The electron's kinetic energy is $K = 136$ eV, so it is not relativistic.

VP39.2.4. **IDENTIFY:** This problem involves the de Broglie wavelength of a proton.

SET UP: $\lambda = h/p$, $K = p^2/2m$, $K = eV$.

Chapter 39 of Student's Solutions Manual to accompany *University Physics with Modern Physics, Volume 3,* Fifteenth Edition.
Hugh D. Young and Roger A. Freedman.

EXECUTE: **(a)** We want the de Broglie wavelength. Using the given speed gives $\lambda = h/mv = 1.67$ pm.

(b) We want the accelerating voltage V. $eV = K = \frac{1}{2}mv^2$. Solving for V using the known quantities gives $V = 295$ V.

EVALUATE: If a proton and electron have comparable speeds, the proton has a *much smaller* de Broglie wavelength because it is much more massive than the electron.

VP39.6.1. **IDENTIFY:** The problem involves the energy due to electron transitions in a hypothetical atom.
SET UP and EXECUTE: We want the wavelengths of the emitted light in each case. The possible transitions are $5 \rightarrow 0$ and $5 \rightarrow 2 \rightarrow 0$, and $\Delta E = hc/\lambda$.
$\underline{5 \rightarrow 0}$: $\Delta E = 5.00$ eV $- 0 = 5.00$ eV. $\lambda = hc/\Delta E = hc/(5.00$ eV$) = 248$ nm.
$\underline{5 \rightarrow 2}$: $\Delta E = 5.00$ eV $- 2.00 = 3.00$ eV. $\lambda = hc/\Delta E = hc/(3.00$ eV$) = 414$ nm.
$\underline{2 \rightarrow 0}$: $\Delta E = 2.00$ eV $- 0 = 2.00$ eV. $\lambda = hc/\Delta E = hc/(2.00$ eV$) = 620$ nm.
EVALUATE: The wavelengths emitted by a gas of this atom would be 248 nm, 414 nm, and 620 nm.

VP39.6.2. **IDENTIFY:** The problem involves the energy due to electron transitions in a hypothetical atom.
SET UP: $\Delta E = hc/\lambda$.
EXECUTE: **(a)** $\Delta E_{1 \rightarrow \text{grd}} = hc/\lambda_1 = hc/(385$ nm$) = 3.22$ eV. Therefore $E_1 = 3.22$ eV relative to ground.
$\Delta E_{2 \rightarrow 1} = hc/\lambda_2 = hc/(674$ nm$) = 1.84$ eV. Therefore $E_2 = 1.84$ eV relative to E_1. Relative to ground we have $E_2 = 1.84$ eV $+ 3.22$ eV $= 5.06$ eV.

(b) $\lambda = hc/\Delta E = hc/(5.06$ eV$) = 245$ nm.
EVALUATE: Note that the energy difference between adjacent levels gets smaller for higher and higher levels.

VP39.6.3. **IDENTIFY:** The problem involves electron transitions in a Bohr hydrogen atom.
SET UP: $\Delta E = hc/\lambda$. For the Bohr hydrogen atom, $E_n = -(13.60$ eV$)/n^2$. We want the energy and wavelength of the emitted photon.
EXECUTE: **(a)** $5 \rightarrow 3$: $\Delta E = (-13.60$ eV$)(1/5^2 - 1/3^2) = 0.967$ eV. $\lambda = hc/\Delta E = hc/(0.967$ eV$)$ $= 1.28$ μm.
(b) $4 \rightarrow 2$: $\Delta E = (-13.60$ eV$)(1/4^2 - 1/2^2) = 2.55$ eV. $\lambda = hc/\Delta E = hc/(2.55$ eV$) = 487$ nm.
(c) $3 \rightarrow 1$: $\Delta E = (-13.60$ eV$)(1/3^2 - 1/1^2) = 12.1$ eV. $\lambda = hc/\Delta E = hc/(12.1$ eV$) = 103$ nm.
EVALUATE: Note that as the energy difference increases, the wavelength of the emitted photon decreases. This is reasonable because shorter wavelength photons have more energy than long wavelength photons.

VP39.6.4. **IDENTIFY:** The problem involves the energy due to electron transitions in a hydrogen atom.
SET UP: $\Delta E = hc/\lambda$. For the Bohr hydrogen atom, $E_n = -(13.60$ eV$)/n^2$, $K_n = (13.60$ eV$)/n^2$, and $U_n = -(27.20$ eV$)/n^2$.
EXECUTE: **(a)** We want the difference in kinetic energy. $\Delta E = K_6 - K_2 = (13.60$ eV$)(1/6^2 - 1/2^2) =$ -3.02 eV.
(b) We want the difference in potential energy.
$\Delta U = U_6 - U_2 = (-27.20$ eV$)(1/6^2 - 1/2^2) = +3.02$ eV.
(c) We want the wavelength of the photon. $\Delta E = hc/\lambda$ gives $\lambda = hc/\Delta E = hc/(6.04$ eV $- 3.02$ eV$)$ $= 411$ nm.
EVALUATE: Check: $\Delta E = K_6 - K_2 = (-13.60$ eV$)(1/6^2 - 1/2^2) = +3.02$ eV, as we used in part (c).

VP39.8.1. **IDENTIFY:** This problem involves blackbody radiation and the Wien law.

SET UP: $I = \sigma T^4$, Wien law: $\lambda_m = \dfrac{2.90 \times 10^{-3} \text{ m} \cdot \text{K}}{T}$.

EXECUTE: **(a)** We want the peak wavelength. Using $T = 3590$ K, the Wien law gives 808 nm. This wavelength is greater than that of visible light, so it is in the *infrared* region.
(b) We want the intensity I. Using $T = 3590$ K, $I = \sigma T^4$ gives 9.42×10^6 W/m^2.
EVALUATE: Betelgeuse is a red giant. It is red because it radiates most of its visible light in the red end of spectrum.

VP39.8.2. **IDENTIFY:** This problem involves blackbody radiation and the Wien law.

SET UP: $I = \sigma T^4$, Wien law: $\lambda_m = \dfrac{2.90 \times 10^{-3} \text{ m} \cdot \text{K}}{T}$.

EXECUTE: **(a)** We want the temperature. Solving $I = \sigma T^4$ for T and using $I = 78.0$ MW/m^2 gives $T = 6090$ K.
(b) We want the peak wavelength. Using $T = 6090$ in the Wien law gives $\lambda_m = 476$ nm.
(c) This wavelength is in the visible (to humans) part of the electromagnetic spectrum.
EVALUATE: This blackbody would be bluish because the peak wavelength is toward the blue end of the spectrum.

VP39.8.3. **IDENTIFY:** This problem involves blackbody radiation and the Wien law.

SET UP: $I = \sigma T^4$, Wien law: $\lambda_m = \dfrac{2.90 \times 10^{-3} \text{ m} \cdot \text{K}}{T}$.

EXECUTE: **(a)** We want the temperature. Using the Wien law with the peak wavelength at 239 nm gives $T = 12{,}100$ K.
(b) We want the power per unit area. Using $I = \sigma T^4$ with $T = 12{,}100$ K gives $I = 1.23 \times 10^9$ W/m^2 = 1.23 GW/m^2.
EVALUATE: Rigel is a very hot star. Its peak wavelength is in the ultraviolet part of the electro-magnetic spectrum.

VP39.8.4 **IDENTIFY:** This problem involves blackbody radiation, the Planck radiation law, and the Wien law.

SET UP: $I = \sigma T^4$, Planck law: $I(\lambda) = \dfrac{2\pi h c^2}{\lambda^5 \left(e^{hc/\lambda kT} - 1 \right)}$, Wien law: $\lambda_m = \dfrac{2.90 \times 10^{-3} \text{ m} \cdot \text{K}}{T}$.

EXECUTE: **(a)** We want the peak wavelength. Using the Wien law with $T = 3040$ K gives $\lambda_m = 954$ nm.
(b) We want I. The intensity within the range $\Delta\lambda$ is $I \approx I(\lambda)\Delta\lambda$ if $\Delta\lambda$ is small, as it is in this case. Using the Planck law gives

$$I = I(\lambda)\Delta\lambda = \dfrac{2\pi h c^2}{\lambda^5 \left(e^{hc/\lambda kT} - 1 \right)} \Delta\lambda.$$

Using $\lambda = 954$ nm, $\Delta\lambda = 12.0$ nm, and $T = 3040$ K gives $I = 40.1$ kW/m^2.
EVALUATE: The total intensity the star radiates is $I = \sigma T^4 = 4.84$ MW/m^2 at $T = 3040$ K. So the fraction in the 12 nm range is only (40.1 kW/m^2)/(4.84 MW/m^2) = 0.83%.

39.3. **IDENTIFY:** For a particle with mass, $\lambda = \dfrac{h}{p}$ and $K = \dfrac{p^2}{2m}$.

SET UP: $1 \text{ eV} = 1.60 \times 10^{-19}$ J.

EXECUTE: **(a)** $\lambda = \dfrac{h}{p} \Rightarrow p = \dfrac{h}{\lambda} = \dfrac{(6.63 \times 10^{-34} \text{ J} \cdot \text{s})}{(2.80 \times 10^{-10} \text{ m})} = 2.37 \times 10^{-24} \text{ kg} \cdot \text{m/s}.$

(b) $K = \dfrac{p^2}{2m} = \dfrac{(2.37 \times 10^{-24} \text{ kg} \cdot \text{m/s})^2}{2(9.11 \times 10^{-31} \text{ kg})} = 3.08 \times 10^{-18} \text{ J} = 19.3 \text{ eV}.$

EVALUATE: This wavelength is on the order of the size of an atom. This energy is on the order of the energy of an electron in an atom.

39.5. **IDENTIFY** and **SET UP:** The de Broglie wavelength is $\lambda = \dfrac{h}{p} = \dfrac{h}{mv}.$

EXECUTE: The de Broglie wavelength is the same for the proton and the electron, so

$\dfrac{h}{m_e v_e} = \dfrac{h}{m_p v_p}.$

$v_p = v_e(m_e/m_p) = (8.00 \times 10^6 \text{ m/s})[(9.109 \times 10^{-31} \text{ kg})/(1.6726 \times 10^{-27} \text{ kg})] = 4360 \text{ m/s} = 4.36 \text{ km/s}.$

EVALUATE: The proton and electron have the same de Broglie wavelength and the same momentum, but very different speeds because $m_p \gg m_e.$

39.11. **IDENTIFY:** The acceleration gives momentum to the electrons. We can use this momentum to calculate their de Broglie wavelength.

SET UP: The kinetic energy K of the electron is related to the accelerating voltage V by $K = eV.$

For an electron $E = \frac{1}{2}mv^2 = \dfrac{p^2}{2m}$ and $\lambda = \dfrac{h}{p}.$ For a photon $E = \dfrac{hc}{\lambda}.$

EXECUTE: **(a)** For an electron $p = \dfrac{h}{\lambda} = \dfrac{6.63 \times 10^{-34} \text{ J} \cdot \text{s}}{5.00 \times 10^{-9} \text{ m}} = 1.33 \times 10^{-25} \text{ kg} \cdot \text{m/s}$ and

$E = \dfrac{p^2}{2m} = \dfrac{(1.33 \times 10^{-25} \text{ kg} \cdot \text{m/s})^2}{2(9.11 \times 10^{-31} \text{ kg})} = 9.71 \times 10^{-21} \text{ J.}$ $V = \dfrac{K}{e} = \dfrac{9.71 \times 10^{-21} \text{ J}}{1.60 \times 10^{-19} \text{ C}} = 0.0607 \text{ V}.$ The elec-

trons would have kinetic energy 0.0607 eV.

(b) $E = \dfrac{hc}{\lambda} = \dfrac{1.24 \times 10^{-6} \text{ eV} \cdot \text{m}}{5.00 \times 10^{-9} \text{ m}} = 248 \text{ eV}.$

(c) $E = 9.71 \times 10^{-21} \text{ J}$, so $\lambda = \dfrac{hc}{E} = \dfrac{(6.63 \times 10^{-34} \text{ J} \cdot \text{s})(3.00 \times 10^8 \text{ m/s})}{9.71 \times 10^{-21} \text{ J}} = 20.5 \text{ } \mu\text{m}.$

EVALUATE: If they have the same wavelength, the photon has vastly more energy than the electron.

39.13. **IDENTIFY:** The intensity maxima are located by $d \sin \theta = m\lambda.$ Use $\lambda = \dfrac{h}{p}$ for the wavelength of the neutrons. For a particle, $p = \sqrt{2mE}.$

SET UP: For a neutron, $m = 1.675 \times 10^{-27} \text{ kg}.$

EXECUTE: For $m = 1$, $\lambda = d \sin \theta = \dfrac{h}{\sqrt{2mE}}.$

$E = \dfrac{h^2}{2md^2 \sin^2 \theta} = \dfrac{(6.63 \times 10^{-34} \text{ J} \cdot \text{s})^2}{2(1.675 \times 10^{-27} \text{ kg})(9.10 \times 10^{-11} \text{ m})^2 \sin^2(28.6°)} = 6.91 \times 10^{-20} \text{ J} = 0.432 \text{ eV}.$

EVALUATE: The neutrons have $\lambda = 0.0436 \text{ nm}$, comparable to the atomic spacing.

39.17. **(a) IDENTIFY:** If the particles are treated as point charges, $U = \dfrac{1}{4\pi\epsilon_0}\dfrac{q_1 q_2}{r}$.

SET UP: $q_1 = 2e$ (alpha particle); $q_2 = 82e$ (lead nucleus); r is given so we can solve for U.

EXECUTE: $U = (8.987\times10^9 \text{ N}\cdot\text{m}^2/\text{C}^2)\dfrac{(2)(82)(1.602\times10^{-19}\text{ C})^2}{6.50\times10^{-14}\text{ m}} = 5.82\times10^{-13}\text{ J}$

$U = 5.82\times10^{-13}\text{ J } (1\text{ eV}/1.602\times10^{-19}\text{ J}) = 3.63\times10^6\text{ eV} = 3.63\text{ MeV}$

(b) IDENTIFY: Apply conservation of energy: $K_1 + U_1 = K_2 + U_2$.

SET UP: Let point 1 be the initial position of the alpha particle and point 2 be where the alpha particle momentarily comes to rest. Alpha particle is initially far from the lead nucleus implies $r_1 \approx \infty$ and $U_1 = 0$. Alpha particle stops implies $K_2 = 0$.

EXECUTE: Conservation of energy thus says $K_1 = U_2 = 5.82\times10^{-13}\text{ J} = 3.63\text{ MeV}$.

(c) $K = \frac{1}{2}mv^2$, so $v = \sqrt{\dfrac{2K}{m}} = \sqrt{\dfrac{2(5.82\times10^{-13}\text{ J})}{6.64\times10^{-27}\text{ kg}}} = 1.32\times10^7\text{ m/s}$.

EVALUATE: $v/c = 0.044$, so it is ok to use the nonrelativistic expression to relate K and v. When the alpha particle stops, all its initial kinetic energy has been converted to electrostatic potential energy.

39.19. **IDENTIFY** and **SET UP:** Use the energy to calculate n for this state. Then use the Bohr equation, $L = n\hbar$, to calculate L.

EXECUTE: $E_n = -(13.6\text{ eV})/n^2$, so this state has $n = \sqrt{13.6/1.51} = 3$. In the Bohr model, $L = n\hbar$, so for this state $L = 3\hbar = 3.16\times10^{-34}\text{ kg}\cdot\text{m}^2/\text{s}$.

EVALUATE: We will find in Section 41.1 that the modern quantum mechanical description gives a different result.

39.21. **IDENTIFY:** The force between the electron and the nucleus in Be^{3+} is $F = \dfrac{1}{4\pi\epsilon_0}\dfrac{Ze^2}{r^2}$, where

$Z = 4$ is the nuclear charge. All the equations for the hydrogen atom apply to Be^{3+} if we replace e^2 by Ze^2.

(a) SET UP: Modify the energy equation for hydrogen, $E_n = -\dfrac{1}{\epsilon_0^2}\dfrac{me^4}{8n^2h^2}$ by replacing e^2 with Ze^2.

EXECUTE: $E_n = -\dfrac{1}{\epsilon_0^2}\dfrac{me^4}{8n^2h^2}$ (hydrogen) becomes

$E_n = -\dfrac{1}{\epsilon_0^2}\dfrac{m(Ze^2)^2}{8n^2h^2} = Z^2\left(-\dfrac{1}{\epsilon_0^2}\dfrac{me^4}{8n^2h^2}\right) = Z^2\left(-\dfrac{13.60\text{ eV}}{n^2}\right)$ (for Be^{3+}).

The ground-level energy of Be^{3+} is $E_1 = 16\left(-\dfrac{13.60\text{ eV}}{1^2}\right) = -218\text{ eV}$.

EVALUATE: The ground-level energy of Be^{3+} is $Z^2 = 16$ times the ground-level energy of H.

(b) SET UP: The ionization energy is the energy difference between the $n \to \infty$ level energy and the $n = 1$ level energy.

EXECUTE: The $n \to \infty$ level energy is zero, so the ionization energy of Be^{3+} is 218 eV.

EVALUATE: This is 16 times the ionization energy of hydrogen.

(c) SET UP: $\frac{1}{\lambda} = R\left(\frac{1}{n_1^2} - \frac{1}{n_2^2}\right)$ just as for hydrogen but now R has a different value.

EXECUTE: $R_H = \frac{me^4}{8\epsilon_0^2 h^3 c} = 1.097 \times 10^7 \text{ m}^{-1}$ for hydrogen becomes $R_{Be} = Z^2 \frac{me^4}{8\epsilon_0^2 h^3 c}$

$= 16(1.097 \times 10^7 \text{ m}^{-1}) = 1.755 \times 10^8 \text{ m}^{-1}$ for Be^{3+}.

For $n = 2$ to $n = 1, \frac{1}{\lambda} = R_{Be}\left(\frac{1}{1^2} - \frac{1}{2^2}\right) = 3R_{Be}/4.$

$$\lambda = 4/(3R_{Be}) = 4/(3(1.755 \times 10^8 \text{ m}^{-1})) = 7.60 \times 10^{-9} \text{ m} = 7.60 \text{ nm}.$$

EVALUATE: This wavelength is smaller by a factor of 16 compared to the wavelength for the corresponding transition in the hydrogen atom.

(d) SET UP: Modify the Bohr equation for hydrogen, $r_n = \epsilon_0 \frac{n^2 h^2}{\pi m e^2}$, by replacing e^2 with Ze^2.

EXECUTE: $r_n = \epsilon_0 \frac{n^2 h^2}{\pi m (Ze^2)}$ (Be^{3+}).

EVALUATE: For a given n the orbit radius for Be^{3+} is smaller by a factor of $Z = 4$ compared to the corresponding radius for hydrogen.

39.25. **IDENTIFY** and **SET UP:** The ionization threshold is at $E = 0$. The energy of an absorbed photon equals the energy gained by the atom and the energy of an emitted photon equals the energy lost by the atom.

EXECUTE: **(a)** $\Delta E = 0 - (-20 \text{ eV}) = 20 \text{ eV}.$

(b) When the atom in the $n = 1$ level absorbs an 18-eV photon, the final level of the atom is $n = 4$. The possible transitions from $n = 4$ and corresponding photon energies are $n = 4 \rightarrow n = 3, 3 \text{ eV}$; $n = 4 \rightarrow n = 2, 8 \text{ eV}$; $n = 4 \rightarrow n = 1, 18 \text{ eV}$. Once the atom has gone to the $n = 3$ level, the following transitions can occur: $n = 3 \rightarrow n = 2, 5 \text{ eV}$; $n = 3 \rightarrow n = 1, 15 \text{ eV}$. Once the atom has gone to the $n = 2$ level, the following transition can occur: $n = 2 \rightarrow n = 1, 10 \text{ eV}$. The possible energies of emitted photons are: 3 eV, 5 eV, 8 eV, 10 eV, 15 eV, and 18 eV.

(c) There is no energy level 8 eV higher in energy than the ground state, so the photon cannot be absorbed.

(d) The photon energies for $n = 3 \rightarrow n = 2$ and for $n = 3 \rightarrow n = 1$ are 5 eV and 15 eV. The photon energy for $n = 4 \rightarrow n = 3$ is 3 eV. The work function must have a value between 3 eV and 5 eV.

EVALUATE: The atom has discrete energy levels, so the energies of emitted or absorbed photons have only certain discrete energies.

39.27. **IDENTIFY** and **SET UP:** The wavelength of the photon is related to the transition energy $E_i - E_f$ of the atom by $E_i - E_f = \frac{hc}{\lambda}$ where $hc = 1.240 \times 10^{-6} \text{ eV} \cdot \text{m}$.

EXECUTE: **(a)** The minimum energy to ionize an atom is when the upper state in the transition has

$E = 0$, so $E_1 = -17.50 \text{ eV}$. For $n = 5 \rightarrow n = 1, \lambda = 73.86 \text{ nm}$ and $E_5 - E_1 = \frac{1.240 \times 10^{-6} \text{ eV} \cdot \text{m}}{73.86 \times 10^{-9} \text{ m}}$

$= 16.79 \text{ eV}. \ E_5 = -17.50 \text{ eV} + 16.79 \text{ eV} = -0.71 \text{ eV}.$ For $n = 4 \rightarrow n = 1, \lambda = 75.63 \text{ nm}$ and

$E_4 = -1.10$ eV. For $n = 3 \rightarrow n = 1$, $\lambda = 79.76$ nm and $E_3 = -1.95$ eV. For $n = 2 \rightarrow n = 1$, $\lambda = 94.54$ nm and $E_2 = -4.38$ eV.

(b) $E_i - E_f = E_4 - E_2 = -1.10$ eV $- (-4.38$ eV$) = 3.28$ eV and $\lambda = \dfrac{hc}{E_i - E_f} = \dfrac{1.240 \times 10^{-6} \text{ eV} \cdot \text{m}}{3.28 \text{ eV}}$

$= 378$ nm.

EVALUATE: The $n = 4 \rightarrow n = 2$ transition energy is smaller than the $n = 4 \rightarrow n = 1$ transition energy so the wavelength is longer. In fact, this wavelength is longer than for any transition that ends in the $n = 1$ state.

39.29. **IDENTIFY:** Apply conservation of energy to the system of atom and photon.

SET UP: The energy of a photon is $E_\gamma = \dfrac{hc}{\lambda}$.

EXECUTE: **(a)** $E_\gamma = \dfrac{hc}{\lambda} = \dfrac{(6.63 \times 10^{-34} \text{ J} \cdot \text{s})(3.00 \times 10^8 \text{ m/s})}{8.60 \times 10^{-7} \text{ m}} = 2.31 \times 10^{-19}$ J $= 1.44$ eV. So the internal energy of the atom increases by 1.44 eV to $E = -6.52$ eV $+ 1.44$ eV $= -5.08$ eV.

(b) $E_\gamma = \dfrac{hc}{\lambda} = \dfrac{(6.63 \times 10^{-34} \text{ J} \cdot \text{s})(3.00 \times 10^8 \text{ m/s})}{4.20 \times 10^{-7} \text{ m}} = 4.74 \times 10^{-19}$ J $= 2.96$ eV. So the final internal energy of the atom decreases to $E = -2.68$ eV $- 2.96$ eV $= -5.64$ eV.

EVALUATE: When an atom absorbs a photon the energy of the atom increases. When an atom emits a photon the energy of the atom decreases.

39.31. **IDENTIFY:** We know the power of the laser beam, so we know the energy per second that it delivers. The wavelength of the light tells us the energy of each photon, so we can use that to calculate the number of photons delivered per second.

SET UP: The energy of each photon is $E = hf = \dfrac{hc}{\lambda} = \dfrac{1.99 \times 10^{-25} \text{ J} \cdot \text{m}}{\lambda}$. The power is the total energy per second and the total energy E_{tot} is the number of photons N times the energy E of each photon.

EXECUTE: $\lambda = 10.6 \times 10^{-6}$ m, so $E = 1.88 \times 10^{-20}$ J. $P = \dfrac{E_{tot}}{t} = \dfrac{NE}{t}$ so

$$\dfrac{N}{t} = \dfrac{P}{E} = \dfrac{0.100 \times 10^3 \text{ W}}{1.88 \times 10^{-20} \text{ J}} = 5.32 \times 10^{21} \text{ photons/s.}$$

EVALUATE: At over 10^{21} photons per second, we can see why we do not detect individual photons.

39.35. **IDENTIFY:** Apply the equation $\dfrac{n_{ex}}{n_g} = e^{-(E_{ex} - E_g)/kT}$ from the section on the laser.

SET UP: The energy of each of these excited states above the ground state is hc/λ, where λ is the wavelength of the photon emitted in the transition from the excited state to the ground state.

EXECUTE: $\dfrac{n_{2P_{3/2}}}{n_{2P_{1/2}}} = e^{-(E_{2P3/2} - E_{2P1/2})/KT}$. From the diagram

$$\Delta E_{3/2-g} = \dfrac{hc}{\lambda_1} = \dfrac{(6.626 \times 10^{-34} \text{ J})(2.998 \times 10^8 \text{ m/s})}{5.890 \times 10^{-7} \text{ m}} = 3.373 \times 10^{-19} \text{ J.}$$

$$\Delta E_{1/2-g} = \frac{hc}{\lambda_2} = \frac{(6.626\times10^{-34}\text{ J})(2.998\times10^8\text{ m/s})}{5.896\times10^{-7}\text{ m}} = 3.369\times10^{-19}\text{ J}.$$

So $\Delta E_{3/2-1/2} = 3.373\times10^{-19}\text{ J} - 3.369\times10^{-19}\text{ J} = 4.00\times10^{-22}\text{ J}.$

$\dfrac{n_{2P_{3/2}}}{n_{2P_{1/2}}} = e^{-(4.00\times10^{-22}\text{ J})/(1.38\times10^{-23}\text{J/K}\cdot500\text{ K})} = 0.944.$ So more atoms are in the $2P_{1/2}$ state.

EVALUATE: At this temperature $kT = 6.9\times10^{-21}$ J. This is greater than the energy separation between the states, so an atom has almost equal probability for being in either state, with only a small preference for the lower energy state.

39.37. **IDENTIFY:** Energy radiates at the rate $H = Ae\sigma T^4$.

SET UP: The surface area of a cylinder of radius r and length l is $A = 2\pi rl$.

EXECUTE: **(a)** $T = \left(\dfrac{H}{Ae\sigma}\right)^{1/4} = \left(\dfrac{100\text{ W}}{2\pi(0.20\times10^{-3}\text{ m})(0.30\text{ m})(0.26)(5.671\times10^{-8}\text{ W/m}^2\cdot\text{K}^4)}\right)^{1/4}.$

$T = 2.06\times10^3$ K.

(b) $\lambda_m T = 2.90\times10^{-3}$ m $\cdot$ K; $\lambda_m = 1410$ nm.

EVALUATE: **(c)** λ_m is in the infrared. The incandescent bulb is not a very efficient source of visible light because much of the emitted radiation is in the infrared.

39.41. **IDENTIFY:** Since the stars radiate as blackbodies, they obey the Stefan-Boltzmann law and Wien's displacement law.

SET UP: The Stefan-Boltzmann law says that the intensity of the radiation is $I = \sigma T^4$, so the total radiated power is $P = \sigma A T^4$. Wien's displacement law tells us that the peak-intensity wavelength is $\lambda_m = (\text{constant})/T$.

EXECUTE: **(a)** The hot and cool stars radiate the same total power, so the Stefan-Boltzmann law gives $\sigma A_h T_h^4 = \sigma A_c T_c^4 \Rightarrow 4\pi R_h^2 T_h^4 = 4\pi R_c^2 T_c^4 = 4\pi(3R_h)^2 T_c^4 \Rightarrow T_h^4 = 9T^4 \Rightarrow T_h = T\sqrt{3} = 1.7T$, rounded to two significant digits.

(b) Using Wien's law, we take the ratio of the wavelengths, giving $\dfrac{\lambda_m(\text{hot})}{\lambda_m(\text{cool})} = \dfrac{T_c}{T_h} = \dfrac{T}{T\sqrt{3}} = \dfrac{1}{\sqrt{3}} = 0.58$, rounded to two significant digits.

EVALUATE: Although the hot star has only 1/9 the surface area of the cool star, its absolute temperature has to be only 1.7 times as great to radiate the same amount of energy.

39.45. **IDENTIFY and SET UP:** The Heisenberg Uncertainty Principle says $\Delta x\Delta p_x \geq \hbar/2$. The minimum allowed $\Delta x\Delta p_x$ is $\hbar/2$. $\Delta p_x = m\Delta v_x$.

EXECUTE: **(a)** $m\Delta x\Delta v_x = \hbar/2.$ $\Delta v_x = \dfrac{\hbar}{2m\Delta x} = \dfrac{1.055\times10^{-34}\text{ J}\cdot\text{s}}{2(1.67\times10^{-27}\text{ kg})(2.0\times10^{-12}\text{ m})} = 1.6\times10^4$ m/s.

(b) $\Delta x = \dfrac{\hbar}{2m\Delta v_x} = \dfrac{1.055\times10^{-34}\text{ J}\cdot\text{s}}{2(9.11\times10^{-31}\text{ kg})(0.250\text{ m/s})} = 2.3\times10^{-4}$ m.

EVALUATE: The smaller Δx is, the larger Δv_x must be.

39.49. **(a)** **IDENTIFY** and **SET UP:** Apply the equation for the reduced mass,

$$m_r = \frac{m_1 m_2}{m_1 + m_2} = \frac{207 m_e m_p}{207 m_e + m_p}, \text{ where } m_e \text{ denotes the electron mass.}$$

EXECUTE: $m_r = \dfrac{207(9.109\times10^{-31}\text{ kg})(1.673\times10^{-27}\text{ kg})}{207(9.109\times10^{-31}\text{ kg}) + 1.673\times10^{-27}\text{ kg}} = 1.69\times10^{-28}\text{ kg.}$

(b) **IDENTIFY:** In the energy equation $E_n = -\dfrac{1}{\epsilon_0^2}\dfrac{me^4}{8n^2 h^2}$, replace $m = m_e$ by m_r: $E_n = -\dfrac{1}{\epsilon_0^2}\dfrac{m_r e^4}{8n^2 h^2}$.

SET UP: Write as $E_n = \left(\dfrac{m_r}{m_H}\right)\left(-\dfrac{1}{\epsilon_0^2}\dfrac{m_H e^4}{8n^2 h^2}\right)$, since we know that $\dfrac{1}{\epsilon_0^2}\dfrac{m_H e^4}{8h^2} = 13.60\text{ eV}$. Here m_H

denotes the reduced mass for the hydrogen atom;

$m_H = (0.99946)(9.109\times10^{-31}\text{ kg}) = 9.104\times10^{-31}\text{ kg.}$

EXECUTE: $E_n = \left(\dfrac{m_r}{m_H}\right)\left(-\dfrac{13.60\text{ eV}}{n^2}\right).$

$$E_1 = \frac{1.69\times10^{-28}\text{ kg}}{9.104\times10^{-31}\text{ kg}}(-13.60\text{ eV}) = 186(-13.60\text{ eV}) = -2.53\text{ keV.}$$

(c) **SET UP:** From part (b), $E_n = \left(\dfrac{m_r}{m_H}\right)\left(-\dfrac{R_H ch}{n^2}\right)$, where $R_H = 1.097\times10^7\text{ m}^{-1}$ is the Rydberg

constant for the hydrogen atom. Use this result in $\dfrac{hc}{\lambda} = E_i - E_f$ to find an expression for $1/\lambda$. The

initial level for the transition is the $n_i = 2$ level and the final level is the $n_f = 1$ level.

EXECUTE: $\dfrac{hc}{\lambda} = \dfrac{m_r}{m_H}\left[-\dfrac{R_H ch}{n_i^2} - \left(-\dfrac{R_H ch}{n_f^2}\right)\right].$

$$\frac{1}{\lambda} = \frac{m_r}{m_H}R_H\left(\frac{1}{n_f^2} - \frac{1}{n_i^2}\right).$$

$$\frac{1}{\lambda} = \frac{1.69\times10^{-28}\text{ kg}}{9.104\times10^{-31}\text{ kg}}(1.097\times10^7\text{ m}^{-1})\left(\frac{1}{1^2} - \frac{1}{2^2}\right) = 1.527\times10^9\text{ m}^{-1}.$$

$\lambda = 0.655\text{ nm.}$

EVALUATE: From Example 39.6, the wavelength of the radiation emitted in this transition in hydrogen is 122 nm. The wavelength for muonium is $\dfrac{m_H}{m_r} = 5.39\times10^{-3}$ times this. The reduced mass for

hydrogen is very close to the electron mass because the electron mass is much less then the proton mass: $m_p/m_e = 1836$. The muon mass is $207 m_e = 1.886\times10^{-28}\text{ kg}$. The proton is only about 10 times more massive than the muon, so the reduced mass is somewhat smaller than the muon mass. The muon–proton atom has much more strongly bound energy levels and much shorter wavelengths in its spectrum than for hydrogen.

39.51. **IDENTIFY:** This problem involves the Pickering emission series and the Bohr atom.

SET UP: $\dfrac{1}{\lambda} = (1.097 \times 10^7 \text{ m}^{-1}) \left[\dfrac{1}{4} - \dfrac{1}{(n/2)^2} \right]$, $n = 5, 6, \ldots$

EXECUTE: (a) We want the longest and shortest wavelengths of the Pickering series. For the longest wavelength (the least energy), the transition is between adjacent levels, so $n = 5$.

$$\dfrac{1}{\lambda} = (1.097 \times 10^7 \text{ m}^{-1}) \left[\dfrac{1}{4} - \dfrac{1}{(5/2)^2} \right]. \quad \lambda = 1013 \text{ nm.}$$

For shortest wavelength, n approaches infinity.

$$\dfrac{1}{\lambda} = (1.097 \times 10^7 \text{ m}^{-1}) \left[\dfrac{1}{4} - \dfrac{1}{(\infty/2)^2} \right]. \quad \lambda = 364.6 \text{ nm.}$$

(b) Shortest wavelength: The transition is between n_L and infinity. The Bohr energy is

$$\Delta E = E_\infty - E_{n_L}. \quad \dfrac{hc}{\lambda_{\min}} = \dfrac{Z^2 E_1}{n^2}. \quad E_1 = 13.60 \text{ eV} \text{ and } \lambda_{\min} = 364.6 \text{ nm} \text{ which gives } Z^2/n^2 = 0.250. \text{ Try}$$

values of n to find a Z that satisfies this equation.
$n = 1$: This cannot work because Z must be a whole number.
$n = 2$: This gives $Z = 1$, which is hydrogen which we know is *not* the atom.
$n = 3$: Z is not a whole number.
$n = 4$: This gives $Z = 2$ and ends on level 4, so $n_L = 4$, $Z = 2$ (helium).
EVALUATE: Find n for the $\lambda = 1013 \text{ nm}$ transition using the Bohr model for a $Z = 2$ atom.

$$\Delta E = \dfrac{hc}{\lambda} = Z^2 (13.60 \text{ eV}) \left(\dfrac{1}{4^2} - \dfrac{1}{n^2} \right).$$

Using $\lambda = 1013 \text{ nm}$ and $Z = 2$ gives $n = 5$, so the transition is from the $n = 5$ state to the $n = 4$ state.

39.53. (a) IDENTIFY and **SET UP:** The photon energy is given to the electron in the atom. Some of this energy overcomes the binding energy of the atom and what is left appears as kinetic energy of the free electron. Apply $hf = E_f - E_i$, the energy given to the electron in the atom when a photon is absorbed.

EXECUTE: The energy of one photon is $\dfrac{hc}{\lambda} = \dfrac{(6.626 \times 10^{-34} \text{ J} \cdot \text{s})(2.998 \times 10^8 \text{ m/s})}{85.5 \times 10^{-9} \text{ m}}$.

$$\dfrac{hc}{\lambda} = 2.323 \times 10^{-18} \text{ J}(1 \text{ eV}/1.602 \times 10^{-19} \text{ J}) = 14.50 \text{ eV.}$$

The final energy of the electron is $E_f = E_i + hf$. In the ground state of the hydrogen atom the energy of the electron is $E_i = -13.60 \text{ eV}$. Thus $E_f = -13.60 \text{ eV} + 14.50 \text{ eV} = 0.90 \text{ eV}$.

EVALUATE: (b) At thermal equilibrium a few atoms will be in the $n = 2$ excited levels, which have an energy of $-13.6 \text{ eV}/4 = -3.40 \text{ eV}, 10.2 \text{ eV}$ greater than the energy of the ground state. If an electron with $E = -3.40 \text{ eV}$ gains 14.5 eV from the absorbed photon, it will end up with $14.5 \text{ eV} - 3.4 \text{ eV} = 11.1 \text{ eV}$ of kinetic energy.

39.55. **IDENTIFY:** Assuming that Betelgeuse radiates like a perfect blackbody, Wien's displacement and the Stefan-Boltzmann law apply to its radiation.

SET UP: Wien's displacement law is $\lambda_{\text{peak}} = \dfrac{2.90 \times 10^{-3} \text{ m} \cdot \text{K}}{T}$, and the Stefan-Boltzmann law

says that the intensity of the radiation is $I = \sigma T^4$, so the total radiated power is $P = \sigma A T^4$.
EXECUTE: **(a)** First use Wien's law to find the peak wavelength:
$$\lambda_m = (2.90 \times 10^{-3} \text{ m} \cdot \text{K})/(3000 \text{ K}) = 9.667 \times 10^{-7} \text{ m}.$$

Call N the number of photons/second radiated. $N \times (\text{energy per photon}) = IA = \sigma A T^4$.

$$N(hc/\lambda_m) = \sigma A T^4. \quad N = \frac{\lambda_m \sigma A T^4}{hc}.$$

$$N = \frac{(9.667 \times 10^{-7} \text{ m})(5.67 \times 10^{-8} \text{ W/m}^2 \cdot \text{K}^4)(4\pi)(600 \times 6.96 \times 10^8 \text{ m})^2 (3000 \text{ K})^4}{(6.626 \times 10^{-34} \text{ J} \cdot \text{s})(3.00 \times 10^8 \text{ m/s})}.$$

$$N = 5 \times 10^{49} \text{ photons/s.}$$

(b) $\dfrac{I_B A_B}{I_S A_S} = \dfrac{\sigma A_B T_B^4}{\sigma A_S T_S^4} = \dfrac{4\pi R_B^2 T_B^4}{4\pi R_S^2 T_S^4} = \left(\dfrac{600 R_S}{R_S}\right)^2 \left(\dfrac{3000 \text{ K}}{5800 \text{ K}}\right)^4 = 3 \times 10^4.$

EVALUATE: Betelgeuse radiates 30,000 times as much energy per second as does our sun!

39.57. **IDENTIFY:** We are applying quantum principles to our moon.

SET UP and EXECUTE: (a) We want the de Broglie wavelength of the moon. First find the speed.
$v = 2\pi r/t$. Using the orbital radius r from Appendix F and $t = 27.3 \text{ d} = (27.3)(86,400 \text{ s})$, we have
$v = 1.0229 \text{ km/s}$. Now use $\lambda = h/p = h/mv$ with the mass m given in the problem and the speed we
just found, giving $\lambda = 8.81 \times 10^{-60} \text{ m}$.
(b) We want the acceleration of the moon. Apply Newton's second law and universal gravitation to
the moon. We also know that the angular momentum is $L = M_{\text{moon}} vR = mh/2\pi$, so $v = mh/2\pi M_{\text{moon}} R$.
Using these relationships gives

$$\frac{GM_{\text{earth}} M_{\text{moon}}}{R^2} = \frac{M_{\text{moon}} v^2}{R}$$

$$\frac{GM_{\text{earth}}}{R} = v^2 = \left(\frac{mh}{2\pi M_{\text{moon}} R}\right)^2$$

$$R = m^2 \left(\frac{h^2}{4\pi^2 M_{\text{moon}}^2 GM_{\text{earth}}}\right) = m^2 a_{\text{moon}}$$

$$a_{\text{moon}} = \frac{h^2}{4\pi^2 M_{\text{moon}}^2 GM_{\text{earth}}}.$$

(c) We want a_{moon}. Using the numbers given in the problem for the masses, our result from part (b)
gives $a_{\text{moon}} = 5.17 \times 10^{-129} \text{ m m/s}^2$.
(d) We want m. Using $R_m = m^2 a_{\text{moon}}$ and the result from part (c) gives $m = 2.73 \times 10^{68}$.
(e) We want E_0. From $E = -E_0/m$ we get $E_0 = -m^2 E$. Using $E = K + U$ gives

$$E = \frac{1}{2} I\omega^2 - \frac{GM_{\text{moon}} M_{\text{earth}}}{R} = \frac{1}{2} M_{\text{moon}} R^2 \left(\frac{2\pi}{T}\right)^2 - \frac{GM_{\text{moon}} M_{\text{earth}}}{R}.$$

Using $E_0 = -m^2 E$ with the above result and $m = 2.73 \times 10^{68}$, we get $E_0 = -2.81 \times 10^{165} \text{ J}.$

EVALUATE: According to our result, the moon is not even close to its ground state.

39.63. **IDENTIFY:** The electrons behave like waves and produce a double-slit interference pattern after passing through the slits.

SET UP: The first angle at which destructive interference occurs is given by $d\sin\theta = \lambda/2$. The de Broglie wavelength of each of the electrons is $\lambda = h/mv$.

EXECUTE: **(a)** First find the wavelength of the electrons. For the first dark fringe, we have $d\sin\theta = \lambda/2$, which gives $(1.25 \text{ nm})(\sin 18.0°) = \lambda/2$, and $\lambda = 0.7725$ nm. Now solve the de Broglie wavelength equation for the speed of the electron:

$$v = \frac{h}{m\lambda} = \frac{6.626\times10^{-34} \text{ J}\cdot\text{s}}{(9.11\times10^{-31} \text{ kg})(0.7725\times10^{-9} \text{ m})} = 9.42\times10^5 \text{ m/s}$$

which is about 0.3% the speed of light, so they are *nonrelativistic*.

(b) Energy conservation gives $eV = \frac{1}{2}mv^2$ and $V = mv^2/2e$

$$= (9.11\times10^{-31} \text{ kg})(9.42\times10^5 \text{ m/s})^2/[2(1.60\times10^{-19} \text{ C})] = 2.52 \text{ V}.$$

EVALUATE: The de Broglie wavelength of the electrons is comparable to the separation of the slits.

39.65. **IDENTIFY:** Both the electrons and photons behave like waves and exhibit single-slit diffraction after passing through their respective slits.

SET UP: The energy of the photon is $E = hc/\lambda$ and the de Broglie wavelength of the electron is $\lambda = h/mv = h/p$. Destructive interference for a single slit first occurs when $a\sin\theta = \lambda$.

EXECUTE: **(a)** For the photon: $\lambda = hc/E$ and $a\sin\theta = \lambda$. Since the a and θ are the same for the photons and electrons, they must both have the same wavelength. Equating these two expressions for λ gives $a\sin\theta = hc/E$. For the electron, $\lambda = h/p = \dfrac{h}{\sqrt{2mK}}$ and $a \Delta \to 0$, Equating these two expressions for λ gives $a\sin\theta = \dfrac{h}{\sqrt{2mK}}$. Equating the two expressions for $a\sin\theta$ gives

$$hc/E = \frac{h}{\sqrt{2mK}}, \text{ which gives } E = c\sqrt{2mK} = (4.05\times10^{-7} \text{ J}^{1/2})\sqrt{K}.$$

(b) $\dfrac{E}{K} = \dfrac{c\sqrt{2mK}}{K} = \sqrt{\dfrac{2mc^2}{K}}$. Since $v \ll c$, $mc^2 > K$, so the square root is >1. Therefore $E/K > 1$, meaning that the photon has more energy than the electron.

EVALUATE: When a photon and a particle have the same wavelength, the photon has more energy than the particle.

39.71. **(a) IDENTIFY** and **SET UP:** $\Delta x\Delta p_x \geq \hbar/2$. Estimate Δx as $\Delta x \approx 5.0\times10^{-15}$ m.

EXECUTE: Then the minimum allowed Δp_x is

$$\Delta p_x \approx \frac{\hbar}{2\Delta x} = \frac{1.055\times10^{-34} \text{ J}\cdot\text{s}}{2(5.0\times10^{-15} \text{ m})} = 1.1\times10^{-20} \text{ kg}\cdot\text{m/s}.$$

(b) IDENTIFY and **SET UP:** Assume $p \approx 1.1\times10^{-20}$ kg·m/s. Use $E^2 = (mc^2)^2 + (pc)^2$ to calculate E, and then $K = E - mc^2$.

EXECUTE: $E = \sqrt{(mc^2)^2 + (pc)^2}$. $mc^2 = (9.109\times10^{-31} \text{ kg})(2.998\times10^8 \text{ m/s})^2 = 8.187\times10^{-14} \text{ J}$.

$pc = (1.1\times10^{-20} \text{ kg}\cdot\text{m/s})(2.998\times10^8 \text{ m/s}) = 3.165\times10^{-12} \text{ J}.$

$E = \sqrt{(8.187\times10^{-14} \text{ J})^2 + (3.165\times10^{-12} \text{ J})^2} = 3.166\times10^{-12} \text{ J}.$

$K = E - mc^2 = 3.166\times10^{-12} \text{ J} - 8.187\times10^{-14} \text{ J} = 3.084\times10^{-12} \text{ J}\times(1 \text{ eV}/1.602\times10^{-19} \text{ J}) = 19 \text{ MeV}.$

(c) IDENTIFY and SET UP: The Coulomb potential energy for a pair of point charges is given by $U = -kq_1q_2/r$. The proton has charge $+e$ and the electron has charge $-e$.

EXECUTE: $U = -\dfrac{ke^2}{r} = -\dfrac{(8.988\times10^9 \text{ N}\cdot\text{m}^2/\text{C}^2)(1.602\times10^{-19} \text{ C})^2}{5.0\times10^{-15} \text{ m}} = -4.6\times10^{-14} \text{ J} = -0.29 \text{ MeV}.$

EVALUATE: The kinetic energy of the electron required by the uncertainty principle would be much larger than the magnitude of the negative Coulomb potential energy. The total energy of the electron would be large and positive and the electron could not be bound within the nucleus.

39.75. IDENTIFY: Assume both the x rays and electrons are at normal incidence and scatter from the surface plane of the crystal, so the maxima are located by $d\sin\theta = m\lambda$, where d is the separation between adjacent atoms in the surface plane.

SET UP: Let primed variables refer to the electrons. $\lambda' = \dfrac{h}{p'} = \dfrac{h}{\sqrt{2mE'}}$.

EXECUTE: $\sin\theta' = \dfrac{\lambda'}{\lambda}\sin\theta$, and $\lambda' = (h/p') = (h/\sqrt{2mE'})$, and so $\theta' = \arcsin\left(\dfrac{h}{\lambda\sqrt{2mE'}}\sin\theta\right).$

$\theta' = \arcsin\left(\dfrac{(6.63\times10^{-34} \text{ J}\cdot\text{s})(\sin 35.8°)}{(3.00\times10^{-11} \text{ m})\sqrt{2(9.11\times10^{-31} \text{ kg})(4.50\times10^{+3} \text{ eV})(1.60\times10^{-19} \text{ J/eV})}}\right) = 20.9°.$

EVALUATE: The x rays and electrons have different wavelengths and the $m=1$ maxima occur at different angles.

39.77. IDENTIFY: The wave (light or electron matter wave) having less energy will cause less damage to the virus.

SET UP: For a photon $E_{ph} = \dfrac{hc}{\lambda} = \dfrac{1.24\times10^{-6} \text{ eV}\cdot\text{m}}{\lambda}$. For an electron $E_e = \dfrac{p^2}{2m} = \dfrac{h^2}{2m\lambda^2}.$

EXECUTE: **(a)** $E = \dfrac{hc}{\lambda} = \dfrac{1.24\times10^{-6} \text{ eV}\cdot\text{m}}{5.00\times10^{-9} \text{ m}} = 248 \text{ eV}.$

(b) $E_e = \dfrac{h^2}{2m\lambda^2} = \dfrac{(6.63\times10^{-34} \text{ J}\cdot\text{s})^2}{2(9.11\times10^{-31} \text{ kg})(5.00\times10^{-9} \text{ m})^2} = 9.65\times10^{-21} \text{ J} = 0.0603 \text{ eV}.$

EVALUATE: The electron has much less energy than a photon of the same wavelength and therefore would cause much less damage to the virus.

39.79. (a) IDENTIFY and SET UP: $U = A|x|$. $F_x = -dU/dx$ relates force and potential. The slope of the function $A|x|$ is not continuous at $x = 0$, so we must consider the regions $x > 0$ and $x < 0$ separately.

EXECUTE: For $x > 0, |x| = x$, so $U = Ax$ and $F = -\dfrac{d(Ax)}{dx} = -A$. For

$x < 0, |x| = -x$, so $U = -Ax$ and $F = -\dfrac{d(-Ax)}{dx} = +A$. We can write this result as $F = -A|x|/x$, valid for all x except for $x = 0$.

(b) IDENTIFY and SET UP: Use the uncertainty principle, expressed as $\Delta p \Delta x \approx h$, and as in Problem 39.78 estimate Δp by p and Δx by x. Use this to write the energy E of the particle as a function of x. Find the value of x that gives the minimum E and then find the minimum E.

EXECUTE: $E = K + U = \dfrac{p^2}{2m} + A|x|.$

$$px \approx h, \text{ so } p \approx h/x.$$

Then $E \approx \dfrac{h^2}{2mx^2} + A|x|.$

For $x > 0, E = \dfrac{h^2}{2mx^2} + Ax.$

To find the value of x that gives minimum E set $\dfrac{dE}{dx} = 0.$

$$0 = \frac{-2h^2}{2mx^3} + A.$$

$$x^3 = \frac{h^2}{mA} \text{ and } x = \left(\frac{h^2}{mA}\right)^{\frac{1}{3}}.$$

With this x the minimum E is

$$E = \frac{h^2}{2m}\left(\frac{mA}{h^2}\right)^{2/3} + A\left(\frac{h^2}{mA}\right)^{1/3} = \frac{1}{2}h^{2/3}m^{-1/3}A^{2/3} + h^{2/3}m^{-1/3}A^{2/3}.$$

$$E = \frac{3}{2}\left(\frac{h^2A^2}{m}\right)^{1/3}.$$

EVALUATE: The potential well is shaped like a V. The larger A is, the steeper the slope of U and the smaller the region to which the particle is confined and the greater is its energy. Note that for the x that minimizes E, $2K = U.$

39.81. **IDENTIFY and SET UP:** For hydrogen-like atoms (1 electron and Z protons), the energy levels are $E_n = (-13.60 \text{ eV})Z^2/n^2$, with $n = 1$ for the ground state. The energy of a photon is $E = hc/\lambda.$

EXECUTE: **(a)** The least energy absorbed is between the ground state ($n = 1$) and the $n = 2$ state, which gives the longest wavelength. So $\Delta E_{1 \to 2} = hc/\lambda.$ Using the energy levels for this atom, we have

$$(-13.6 \text{ eV})Z^2\left(\frac{1}{2^2} - \frac{1}{1^2}\right) = \frac{hc}{\lambda} \to (10.20 \text{ eV})Z^2 = hc/\lambda. \text{ Solving } Z \text{ gives}$$

$$Z = \sqrt{\frac{hc}{(10.20 \text{ eV})\lambda}} = \sqrt{\frac{(4.136 \times 10^{-15} \text{ eV} \cdot \text{s})(2.998 \times 10^8 \text{ m/s})}{(10.20 \text{ eV})(13.56 \times 10^{-9} \text{ m})}} = 3.0.$$

(b) The next shortest wavelength is between the $n = 3$ and $n = 1$ states.

$$\Delta E_{1 \to 3} = (-13.6 \text{ eV})(3)^2\left(\frac{1}{3^2} - \frac{1}{1^2}\right) = \frac{hc}{\lambda}.$$

Solving for λ gives
$\lambda = (4.136 \times 10^{-15} \text{ eV} \cdot \text{s})(2.998 \times 10^8 \text{ m/s})/(108.8 \text{ eV}) = 11.40 \text{ nm}.$

(c) By energy conservation, $E_{photon} = E_{ionization} + K_{el}$. The ionization energy is the minimum energy to completely remove an electron from the atom, which is from the $n = 1$ state to the $n = \infty$ state.

Therefore $E_{ionization} = (13.60 \text{ eV})Z^2 = (13.60 \text{ eV})(9)$. Therefore the kinetic energy of the electron is
$$K_{el} = E_{photon} - E_{ionization} = hc/\lambda - E_{ionization}.$$
$K_{el} = (4.136 \times 10^{-15} \text{ eV} \cdot \text{s})(2.998 \times 10^8 \text{ m/s})/(6.78 \times 10^{-9} \text{ m}) - (13.60 \text{ eV})(9) = 60.5 \text{ eV}.$

EVALUATE: The energy levels for a $Z = 3$ atom are 9 times as great as for the comparable energy levels in hydrogen, so the wavelengths of the absorbed light are much shorter than they would be for comparable transitions in hydrogen.

39.83. **IDENTIFY and SET UP:** The power radiated by an ideal blackbody is $P = \sigma AT^4$. Wien's displacement law, $\lambda_m T = 2.90 \times 10^{-3}$ m · K, applies to the stars. The surface area of a star is $A = 4\pi R^2$, and $R_{sun} = 6.96 \times 10^8$ m.

EXECUTE: (a) Calculate the radiated power for each star using $P = \sigma AT^4$. For Polaris we have
$$P = \sigma AT^4 = (5.67 \times 10^{-8} \text{ W/m}^2 \cdot \text{K}^4)(4\pi)[(46)(6.96 \times 10^8 \text{ m})]^2 (6015 \text{ K})^4 = 9.56 \times 10^{29} \text{ W}.$$
Repeating this calculation for the other stars gives us the following results.
Polaris: $P = 9.56 \times 10^{29}$ W
Vega: $P = 2.19 \times 10^{28}$ W
Antares: $P = 3.60 \times 10^{31}$ W
α Centauri B: $P = 1.98 \times 10^{26}$ W
Antares has the greatest radiated power.

(b) Apply Wien's displacement law, $\lambda_m T = 2.90 \times 10^{-3}$ m · K, and solve for λ_m. For example, for Polaris we have $\lambda_m = \dfrac{2.90 \times 10^{-3} \text{ m} \cdot \text{K}}{T} = \dfrac{2.90 \times 10^{-3} \text{ m} \cdot \text{K}}{6015 \text{ K}} = 4.82 \times 10^{-7}$ m $= 482$ nm. Repeating this calculation for the other stars gives the following results.
Polaris: $\lambda_m = 482$ nm
Vega: $\lambda_m = 302$ nm
Antares: $\lambda_m = 853$ nm
α Centauri B: $\lambda_m = 551$ nm
The visible range is 380 nm to 750 nm, so Polaris and α Centauri B radiate chiefly in the visible range.

(c) By comparing the results in part (a), we see that only α Centauri B radiates less than our sun.

EVALUATE: The power radiated by a star depends on its surface area *and* its surface temperature. Vega, a very hot star, radiates less than the much cooler Antares because Antares has over 300 times the radius of Vega and therefore over 300^2 times the surface area of Vega. A hot star is not necessarily brighter than a cool star.

40

STUDY GUIDE FOR QUANTUM MECHANICS

Summary

In this chapter, our particle investigation turns to an analysis of wave functions—an analysis that will be used to predict the probability of a particle being in a specific location at a specific time. We'll explore the quantum mechanics of particles trapped in bound states, such as electrons orbiting atoms. We'll apply the Schrödinger equation to find wave function solutions of it in a variety of cases, including a particle confined to a box, a particle in a square well, and a particle in a harmonic oscillator potential. We'll also investigate phenomena forbidden by Newtonian mechanics, such as quantum mechanical tunneling.

Objectives

After studying this chapter, you'll understand

- How particles are described in terms of wave functions.
- The use of the Schrödinger equation to determine the behavior of particles.
- How to calculate the wave functions for a particle in a box.
- How to determine the wave function for a particle in a potential well.
- The definitions of *tunneling* and *barrier penetration*.
- How to solve the problem of a particle in a harmonic oscillator potential.

Concepts and Equations

Term	Description
Wave Functions	The wave function $\Psi(x, y, z, t)$ for a particle contains all of the information about the particle. The quantity $\left\lvert\Psi(x, y, z, t)\right\rvert^2$ is the probability distribution function that determines the relative probability of finding a particle near a given position at a given time. For a particle in a definite energy state, called a stationary state, the wave function can be separated into a spatial component and a temporal component: $$\Psi(x, y, z, t) = \psi(x, y, z)\, e^{-iEt/\hbar}.$$
The Schrödinger Equation	The Schrödinger equation can be used to determine the wave function for a particle moving in one dimension in the presence of a potential-energy function $U(x)$. The Schrödinger equation is $$-\frac{\hbar^2}{2m}\frac{d^2\psi(x)}{dx^2} + U(x)\psi(x) = e\psi(x).$$
Particle in a Box	The energy levels for a particle of mass m in an infinitely deep square well potential of width L are given by $$E_n = \frac{p_n^2}{2m} = \frac{n^2 h^2}{8mL^2}, \qquad n = 1, 2, 3, \dots.$$ The corresponding normalized particle wave functions are given by $$\psi_n(x) = \sqrt{\frac{2}{L}}\,\sin\frac{n\pi x}{L}, \qquad n = 1, 2, 3, \dots.$$
Wave Functions and Normalization	To be a solution of the Schrödinger equation, the wave function and its derivative must be continuous everywhere. Wave functions are usually normalized such that the probability of finding the particle somewhere is unity: $$\int_{-\infty}^{\infty}\left\lvert\psi(x)\right\rvert^2\,dx = 1$$
Finite Potential Well	The energy levels in a potential well of finite depth are lower than those in an infinite well. The levels are obtained by matching wave functions at the well walls and satisfying continuity of the wave function and its derivative.
Potential Barriers and Tunneling	Because of a process called tunneling, there is a finite, nonzero probability that a particle will penetrate a potential-energy barrier even if its initial kinetic energy is less than the height of the barrier.
Quantum Harmonic Oscillator	The energy levels for a harmonic oscillator for which $U(x) = 1/2\,k'x^2$ are $$E_n = \left(n + \frac{1}{2}\right)\hbar\omega = \left(n + \frac{1}{2}\right)\hbar\sqrt{\frac{k'}{m'}}, \qquad n = 0, 1, 2, \dots.$$

Key Concept 1: A free particle (one on which no forces act) that moves in the x-direction only is described by a wave function $\Psi(x, t)$ that obeys the free-particle Schrödinger equation [Eq. (40.15)]. A wave function with definite momentum and energy is completely delocalized; to write a wave function that is localized in space, you must superpose wave functions with different values of momentum and energy.

Key Concept 2: A quantum-mechanical state of definite energy E has a wave function of the form $\Psi(x, t) = \psi(x)e^{-iEt/\hbar}$. It is called a stationary state because the associated probability distribution function $\left\lvert\Psi(x, t)\right\rvert^2 = \left\lvert\psi(x)\right\rvert^2$ does not depend on time.

Key Concept 3: The wave functions for the stationary states of a particle confined to a one-dimensional box are sinusoidal standing waves inside the box with a node at each end; the wave function is zero outside the box. The nth stationary state has a wave function with n half-wavelengths within the length of the box, and has an energy equal to n^2 times the energy of the lowest-energy $(n=1)$ state. There are an infinite number of these stationary states.

Key Concept 4: To be physically reasonable, a stationary-state wave function for a given situation must satisfy both the Schrödinger equation and any boundary conditions that apply to that situation.

Key Concept 5: A finite potential well is a region where the potential energy is constant and less than the constant value outside that region. If a particle is in a bound state, so that in Newtonian mechanics it would not have enough energy to escape from the well, its stationary-state wave function is sinusoidal within the well and exponential outside the well.

Key Concept 6: A finite square well has a finite number of bound states. The energies of these bound states are determined by the boundary conditions that the wave function and its derivative are both continuous at the edges of the well.

Key Concept 7: Quantum-mechanical particles can be found in places where Newtonian mechanics would forbid them to be. An example is tunneling, in which a particle can pass through a barrier where its total mechanical energy is less than the potential energy.

The probability that tunneling occurs depends on the width of the barrier and the additional energy the particle would need to "climb" over the barrier.

Key Concept 8: As in Newtonian physics, in a quantum-mechanical harmonic oscillator a particle of mass m moves under the influence of the potential-energy function of an ideal spring with force constant k'. The lowest-energy stationary state of the harmonic oscillator has energy $\frac{1}{2}\hbar\omega$, where $\omega = \sqrt{k'/m}$ is the Newtonian angular frequency; each excited state has $\hbar\omega$ more energy than the state below it.

Conceptual Questions

1: Tennis-ball wave function

Tennis balls are made of electrons and other subatomic particles, so a tennis ball can be described in terms of a combination of wave functions. If tennis balls are wave functions, couldn't they exhibit destructive interference and disappear before reaching your racket?

IDENTIFY, SET UP, AND EXECUTE It is true that the particles which make up the tennis ball are described by wave functions. The wave functions describe the particles as spread out in a region of space, and the waves can interfere destructively. However, the wavelength of the tennis ball is so small that any interference effects that might arise are below the threshold of visibility.

EVALUATE As we look at the strange subatomic world, we need to make sure that our interpretations and predictions relating to the macroscopic world remain valid. Quantum mechanics, as we see in this problem, apply to the macroscopic world, but the effects are so small that they do not change our existing interpretation.

2: Probability in a box

Consider the allowed energy states of a particle in a box. How does the probability of finding the particle in the left half of the box compare with the probability of finding the particle in the right half of the box, for any energy state?

IDENTIFY, SET UP, AND EXECUTE The box is symmetric; therefore, the probability of finding a particle in the left half of the box is equal to the probability of finding the particle in the right half of the box, regardless of the energy state.

EVALUATE Building intuition about quantum mechanics and particle probabilities will help you prepare for the same type of concepts that are found throughout the text.

3: Interpreting wave functions

Consider the wave function for a particle in a box, shown in Figure 1. Where are the locations at which the particle is most likely to be found? Where are the locations at which the particle is least likely to be found?

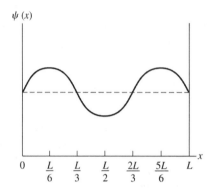

Figure 1 Question 2.

IDENTIFY, SET UP, AND EXECUTE The figure shows the wave function for the $n = 3$ state. The wave function has three nodes and crosses zero twice in the box. The probability function is the square of the wave function. When the wave function is squared, there will be three maxima, located at $L/6$, $L/2$, and $5L/6$. The particle is most likely to be found at these three locations.

There are two minima in the probability function, located at $L/3$ and $2L/3$, corresponding to locations where the particle is least likely to be found. In addition, the particle will not likely be found at the edges of the box: 0 and L.

EVALUATE It is important to remember that the probability function is the *square* of the wave function. Negative values of a wave function do not correspond to locations where the particle is least likely to be found. Here, the minimum of the wave function is one of the most likely places for the particle to be found.

Problems

1: Electron in a well

Suppose an electron is confined to an infinitely deep, one-dimensional potential well of length L. Calculate the value of L required for the frequency of a photon emitted in a transition from the $n = 2$ state to the $n = 1$ state to be equal to the frequency of a photon emitted in a transition from the $n = 2$ state to the $n = 1$ state in the Bohr hydrogen model.

IDENTIFY AND SET UP We'll use energy levels for a particle in a box and the Bohr model to solve the problem.

EXECUTE The energy levels for a particle in an infinitely deep well of length L are given by

$$E_n = \frac{n^2 h^2}{8mL^2}.$$

The energy needed for the transition between the $n=2$ and $n=1$ states is then

$$\Delta E_{\text{box}} = \frac{h^2}{8mL^2}(2^2 - 1^2).$$

Solving for L, we have

$$L = \sqrt{\frac{3h^2}{8m\Delta E}}.$$

We need the energy difference, which we find from the Bohr model. The energy for the n level in the Bohr model is given by

$$E_n = -\frac{hcR}{n^2} = -\frac{13.60 \text{ eV}}{n^2}.$$

Solving for the energy difference gives

$$\Delta E_{\text{H atom}} = E_2 - E_1 = \frac{-13.60 \text{ eV}}{2^2} - \frac{-13.60 \text{ eV}}{1^2} = 10.2 \text{ eV}.$$

Substituting this result into the earlier result allows us to solve for L:

$$L = \sqrt{\frac{3(6.626 \times 10^{-34} \text{ Js})^2}{8(9.1 \times 10^{-31} \text{ kg})(10.2 \text{ eV})(1.6 \times 10^{-19} \text{ J/eV})}} = 3.3 \times 10^{-10} \text{ m}.$$

KEY CONCEPT 3 **EVALUATE** How does this result compare with the circumference of the first Bohr orbit? It approximately matches it, indicating that a square well potential serves as a reasonable model of the hydrogen atom.

Practice Problem: Repeat the problem for $(n=3)$ instead of $(n=2)$ *Answer:* 4.99×10^{-10} m

Extra Practice: What is the length of the well when the $(n=2)$ to $(n=1)$ transition has the same energy as the ionization energy of hydrogen? *Answer:* 2.88×10^{-10} m

2: Finding the wave function

Find the wave function of a particle in a box centered at $x=0$ with walls at $x = \pm L/2$.

IDENTIFY AND SET UP We'll solve the Schrödinger equation to solve the problem. We'll need to check the boundary conditions to ensure that our solution is correct.

EXECUTE The Schrödinger equation for a potential $U=0$ is given by

$$-\frac{\hbar^2}{2m}\frac{d^2\psi}{dx^2} = E\psi.$$

The solutions are given by

$$\psi_1 = \sin kx,$$
$$\psi_2 = \cos kx.$$

We substitute these solutions into the Schrödinger equation and find that

$$E = \frac{\hbar^2 k^2}{2m}.$$

We now check the boundary conditions (i.e., the wave functions go to zero at the walls). For the first wave function, we have

$$\psi_1(-L/2) = \psi_1(L/2) = 0.$$

From this, we see that argument of the sin function must be integer multiples of π, or

$$\frac{kL}{2} = m\pi, \qquad m = 1, 2, \ldots$$

$$k = \frac{2m\pi}{L} = \frac{n\pi}{L} \qquad \text{for even } n.$$

Applying the boundary condition to the second wave function gives

$$\psi_2(-L/2) = \psi_2(L/2) = 0.$$

This limits the values of k to

$$\frac{kL}{2} = \frac{n\pi}{2},$$

$$k = \frac{n\pi}{L} \qquad \text{for odd } n.$$

We use these results to find the energy levels. With the given k's, the energy levels are

$$E_n = \frac{\hbar^2 k^2}{2m} = \frac{h^2 n^2}{8mL^2}.$$

The corresponding wave functions are

$$\psi_n = \sin\frac{n\pi}{L}x \qquad \text{for even } n,$$

$$\psi_n = \cos\frac{n\pi}{L}x \qquad \text{for odd } n.$$

KEY CONCEPT 4 **EVALUATE** We see that the result is similar to the box wave functions between 0 and L, but with the wave functions displaced by $L/2$.

3: Tunneling of a car

A car of mass 3000 kg rolls without friction on a level track and approaches a hill of height 1.0 m and width 1.0 m. It has enough kinetic energy so that it will rise to a height of 0.5 m and then return to the track. What is the probability that the car will tunnel through the hill?

IDENTIFY AND SET UP We'll use the probability-of-tunneling function, equation 42, in the text, to solve the problem.

EXECUTE Given kinetic energy E, the probability of tunneling through a barrier of height U_0 and width L is given by the transmission coefficient, approximated by

$$T = Ge^{-2\kappa L},$$

where

$$G = 16\frac{E}{U_0}\left(1 - \frac{E}{U_0}\right), \quad \kappa = \frac{\sqrt{2m(U_0 - E)}}{\hbar}.$$

For this problem,

$$U_0 = mgY, \quad E = \frac{U_0}{2}.$$

Substituting, we find that

$$G = 16\frac{1}{2}\left(1 - \frac{1}{2}\right) = 4$$

and

$$\kappa = \frac{m\sqrt{gY}}{\hbar}.$$

The exponential factor becomes

$$2\kappa L = \frac{2m\sqrt{gY}L}{\hbar} = \frac{2(3000 \text{ kg})\sqrt{(9.8 \text{ m/s}^2)(1.0 \text{ m})}(1.0 \text{ m})}{1.05 \times 10^{-34} \text{ Js}} = 1.8 \times 10^{38}.$$

The transmission factor is therefore

$$T = 4\frac{1}{e^{1.8 \times 10^{38}}} \ll 1.$$

KEY CONCEPT 7 **EVALUATE** Clearly, this probability is extremely small. It is so small that you would never expect to see the car tunneling through the hill, much as common sense tells us.

Practice Problem: How big does the car need to be to have a one-in-a-billion chance of tunneling through? *Answer:* 3.7×10^{-34} kg

Extra Practice: What is the probability that an electron could tunnel through? *Answer:* $4e^{-54321}$

Try It Yourself!

1: Particle in a well

A particle of mass m is confined by a potential well of depth V_0 and width d. For a given value of d, what is the minimum value of V_0 necessary to confine the particle?

Solution Checkpoints

IDENTIFY AND SET UP Use the uncertainty principle to solve the problem. Draw a sketch of the potential-energy function.

EXECUTE For the particle to be confined to the well, we must set the uncertainty in position to the width. The uncertainty in momentum is then

$$\Delta p_x = \frac{\hbar}{\Delta x} = \frac{\hbar}{d}.$$

This gives a corresponding uncertainty in kinetic energy:

From Chapter 40 of Student's Study Guide to accompany *University Physics with Modern Physics, Volume 3*, Fifteenth Edition. Hugh D. Young and Roger A. Freedman. Copyright © 2020 by Pearson Education, Inc. All rights reserved.

$$\Delta E_k = \frac{(\Delta p_x)^2}{2m} = \frac{\hbar^2}{2md^2}.$$

How much energy must the particle have to be confined? The total energy must be less than zero, or

$$E = \Delta E_k - V_0 < 0.$$

Substituting into this expression gives the values of V_0 necessary for confinement:

$$V_0 > \frac{\hbar^2}{2md^2}.$$

The minimum value of V_0 occurs when it is equal to the expression.

KEY CONCEPT 5   **EVALUATE** What is the minimum depth required to produce a bound state for a proton in a nucleus $\left(d = 3.0 \times 10^{-15} \text{ m}\right)$? 2.3 MeV.

Practice Problem: What is the minimum depth required to produce a bound state for an electron in a hydrogen atom $\left(d = 2a_0\right)$? *Answer:* 3.4 eV

Extra Practice: What depth is required when the electron is in its second excited state? *Answer:* 0.042 eV

2: Electron in a box

An electron trapped in a box has a ground-state energy of 10 eV. How big is the box?

Solution Checkpoints

IDENTIFY AND SET UP Use energy levels for a particle in a box.

EXECUTE The energy levels for a particle in an infinitely deep well of length L are given by

$$E_n = \frac{n^2 h^2}{8mL^2}.$$

The ground state corresponds to $n = 1$. Substituting into the equation gives $L = 1.94 \times 10^{-10}$ m.

KEY CONCEPT 3   **EVALUATE** Does this result seem reasonable? Is it comparable to the size of the hydrogen atom?

Practice Problem: How big is the box if it is a proton trapped inside? *Answer:* 4.53×10^{-12} m

Extra Practice: What is the energy of the third excited state? *Answer:* 160 eV

Key Example Variation Problems

Solutions to these problems are in Chapter 40 of the Student's Solutions Manual.

Be sure to review EXAMPLES 40.3 and 40.4 (Section 40.2) before attempting these problems.

VP40.4.1 The lowest energy level for an electron confined to a one-dimensional box is 2.00×10^{-19} J. Find (a) the width of the box and (b) the energies of the $n = 2$ and $n = 3$ energy levels.

VP40.4.2 A proton (mass 1.67×10^{-27} kg) is confined to a one-dimensional box of width 5.00×10^{-15} m. Find the energy difference between (a) the $n = 2$ and $n = 1$ energy levels and (b) the $n = 3$ and $n = 2$ energy levels.

VP40.4.3 A photon is emitted when an electron in a one-dimensional box transitions from the $n = 2$ energy level to the $n = 1$ energy level. The wavelength of this photon is 655 nm. Find (a) the energy of this photon, (b) the width of the box, and (c) the wavelength of the photon emitted when the electron transitions from the $n = 3$ level to the $n = 2$ level.

VP40.4.4 A wave function for a particle in a box with energy E is

$$\psi(x) = A\cos\left(\frac{x\sqrt{2mE}}{\hbar} + \phi\right)$$

(a) Find the second derivative of $\psi(x)$ and show that this function satisfies the Schrödinger equation. (b) Use the boundary condition at $x = 0$ to determine the value of ϕ. (c) Use the boundary condition at $x = L$ to determine the value of E.

Be sure to review EXAMPLE 40.6 (Section 40.3) before attempting these problems.

VP40.6.1 An electron is confined to a potential well of width 0.350 nm. (a) Find the ground-level energy $E_{1-\text{IDW}}$ if the well is infinitely deep. If instead the depth U_0 of the well is six times the value of $E_{1-\text{IDW}}$, find (b) the ground-level energy and (c) the minimum energy required to free the electron from the well.

VP40.6.2 A finite potential well has a depth U_0 that is six times greater than the ground-level energy $E_{1-\text{IDW}}$ of an electron in an infinitely deep well of the same width. The photon emitted when an electron in the finite potential well transitions from the $n = 3$ level to the $n = 2$ level has energy 2.50×10^{-19} J. Find (a) the value of $E_{1-\text{IDW}}$, (b) the value of U_0, and (c) the width of the potential well.

VP40.6.3 An electron in a finite potential well can emit a photon of only one of three different wavelengths when it makes a transition from one energy level to a lower energy level. The depth of this potential well is six times greater than the ground-level energy $E_{1-\text{IDW}}$ that the electron would have if the well had the same width but was infinitely deep. (a) If the shortest of the three wavelengths is 355 nm, find the quantum numbers n for the initial and final electron energy levels for this transition. (b) Find the other two wavelengths and the quantum numbers n of the initial and final energy levels for each transition.

VP40.6.4 An electron is in a potential well that is 0.400 nm wide. (a) Find the ground-level energy of the electron $E_{1-\text{IDW}}$, in eV, if the potential well is infinitely deep. (b) If the depth of the potential is only 0.015 eV, much less than $E_{1-\text{IDW}}$, there is only one bound state. Find the energy, in eV, of this bound state.

Be sure to review EXAMPLE 40.7 (Section 40.4) before attempting these problems.

VP40.7.1 An electron of energy 3.75 eV encounters a potential barrier 6.10 eV high. Find the probability that the electron will tunnel through the barrier if the barrier width is (a) 0.750 nm and (b) 0.500 nm.

VP40.7.2 A 3.50 eV electron encounters a barrier with height 4.00 eV. Find (a) the probability that the electron will tunnel through the barrier if its width is 0.800 nm and (b) the barrier width that gives twice the tunneling probability you found in part (a).

VP40.7.3 A potential barrier is 5.00 eV high and 0.900 nm wide. Find the probability that an electron will tunnel through this barrier if its energy is (a) 4.00 eV, (b) 4.30 eV, and (c) 4.60 eV.

VP40.7.4 There is a 1-in-417 probability that an electron with energy 3.00 eV will tunnel through a barrier with height 5.00 eV. Find the width of the barrier.

STUDENT'S SOLUTIONS MANUAL FOR QUANTUM MECHANICS

VP40.4.1. **IDENTIFY:** This problem involves an electron in a one-dimensional box.

SET UP: $E_n = n^2 \dfrac{\pi^2 h^2}{8mL^2}$.

EXECUTE: **(a)** We want the width L of the box. Solve the energy equation for L with $n = 1$ and use the given energy for E_1. $L = \dfrac{h}{\sqrt{8mE_1}} = 0.549$ nm.

(b) We want the energy. Use $E_n = n^2 E_1$ with the given energy for E_1.

$E_2 = 2^2 E_1 = 4(2.00 \times 10^{-19} \text{ J}) = 8.00 \times 10^{-19}$ J. $E_3 = 3^2 (2.00 \times 10^{-19} \text{ J}) = 1.80 \times 10^{-19}$ J.

EVALUATE: Note that the energy levels are not evenly spaced.

VP40.4.2. **IDENTIFY:** This problem involves an electron in a one-dimensional box.

SET UP: $E_n = n^2 \dfrac{\pi^2 h^2}{8mL^2}$. We want the energy difference between levels.

EXECUTE: **(a)** Using the energy equation, the energy difference between levels is

$$\Delta E_{2,1} = \frac{h^2}{8mL^2}\left(2^2 - 1^2\right) = \frac{3h^2}{8mL^2} = \frac{3h^2}{8m\left(5.00 \times 10^{-15} \text{ m}\right)^2} = 3.94 \times 10^{-12} \text{ J.}$$

(b) Using the same equation as in part (a) gives

$$\Delta E_{2,1} = \frac{h^2}{8mL^2}\left(3^2 - 2^2\right) = \frac{5h^2}{8mL^2} = 6.57 \times 10^{-12} \text{ J.}$$

EVALUATE: Note that the energy difference between adjacent levels increases as the levels increase to higher values of n.

VP40.4.3. **IDENTIFY:** This problem involves transitions by an electron in a one-dimensional box.

SET UP: $E_n = n^2 \dfrac{\pi^2 h^2}{8mL^2}$, $E = hc/\lambda$.

EXECUTE: **(a)** We want the energy of the photon. $E = hc/\lambda = hc/(655 \text{ nm}) = 3.03 \times 10^{-19}$ J.
(b) We want the length L of the box. The energy of the photon is equal to the energy difference between the $n = 1$ and $n = 2$ levels.

$$E_{\text{ph}} = \frac{h^2}{8mL^2}\left(2^2 - 1^2\right) = \frac{3h^2}{8mL^2}.$$

Solve for L using the photon energy from part (a), giving $L = 0.772$ nm.

(c) We want the wavelength of the photon. The energy of the photon is the energy difference between the $n = 2$ and $n = 3$ levels. Use $E = hc/\lambda$ and $L = 0.772$ nm and solve for λ.

$$E_{\text{ph}} = \frac{h^2}{8mL^2}(3^2 - 2^2) = \frac{5h^2}{8mL^2} = \frac{hc}{\lambda}. \quad \lambda = \frac{8mL^2c}{5h} = 393 \text{ nm}.$$

EVALUATE: Note that the wavelength for the $3 \to 2$ transition is less than the wavelength for the $2 \to 1$ transition. This result is reasonable because the energy difference is greater for the $3 \to 2$ transition than it is for the $2 \to 1$ transition.

VP40.4.4. IDENTIFY: This problem involves the wave function for a particle in a box.
SET UP: The wave function and Schrödinger equation are

$$\psi(x) = A\cos\left(\frac{x\sqrt{2mE}}{\hbar} + \phi\right), \quad -\frac{\hbar^2}{2m}\frac{d^2\psi}{dx} = E\psi.$$

EXECUTE: **(a)** We want to show that the given wave function satisfied the Schrödinger equation. First take the second derivative of the wave function and then multiply it by $-\hbar^2/2m$.

$$\frac{d^2\psi}{dx^2} = -\frac{A(2mE)}{\hbar^2}\cos\left(\frac{x\sqrt{2mE}}{\hbar} + \phi\right) - \frac{\hbar^2}{2m}\frac{d^2\psi}{dx^2} = -\frac{\hbar^2}{2m}\left(-\frac{A(2mE)}{\hbar^2}\right)\cos\left(\frac{x\sqrt{2mE}}{\hbar} + \phi\right)$$

$$= EA\cos\left(\frac{x\sqrt{2mE}}{\hbar} + \phi\right) = E\psi.$$

(b) We want ϕ. The wave function must be zero at $x = 0$, which gives $\cos\phi = 0$, so $\phi = \pm\pi/2$.
(c) We want E. At $x = L$ the wave function must be zero, which gives

$$A\cos\left(\frac{L\sqrt{2mE}}{\hbar} \pm \frac{\pi}{2}\right) = 0. \quad \frac{L\sqrt{2mE}}{\hbar} \pm \frac{\pi}{2} = \frac{\pi}{2}.$$

L cannot be zero, so we must have

$$\frac{L\sqrt{2mE_n}}{\hbar} = n\pi. \quad E_n = n^2\frac{\pi^2\hbar^2}{2mL^2}, n = 1, 2, 3, \cdots$$

EVALUATE: The result in part (c) agrees with Eq. (40.31) in the text.

VP40.6.1. IDENTIFY: We are dealing with an electron in a finite potential well.
SET UP: For an infinitely deep well the energy levels are

$$E_{n\text{-IDW}} = n^2\frac{\pi^2\hbar^2}{2mL^2}$$

and for a finite well we use Figure 40.15 in the textbook.
EXECUTE: **(a)** We want the ground level energy, so $n = 1$.

$$E_{1\text{-IDW}} = \frac{\pi^2\hbar^2}{2mL^2} = \frac{\pi^2\hbar^2}{2m(0.350 \text{ nm})^2} = 4.92\times10^{-19} \text{ J}.$$

(b) We want the ground level energy. In this case, $U_0 = 6E_{1\text{-IDW}}$. From Figure 40.15b in the text, we see that $E_1 = 0.625E_{1\text{-IDW}}$, which gives

$$E_1 = (0.625)(4.92\times10^{-19} \text{ J}) = 3.07\times10^{-19} \text{ J}.$$

(c) We want the minimum energy to free the electron. The electron already has 3.07×10^{-19} J of energy, and to be free it needs to have a minimum of

$$U_0 = 6E_{1\text{-IDW}} = 6(4.92\times10^{-19} \text{ J}) = 2.95\times10^{-18} \text{ J}.$$

So, the additional energy it needs is $U_0 - E_1 = 2.64 \times 10^{-18}$ J.

EVALUATE: If the well were infinite, the electron would need infinite energy to escape, meaning that it could not escape.

VP40.6.2. **IDENTIFY:** We are dealing with particle in a finite potential well.

SET UP: $U_0 = 6E_{1\text{-IDW}}$, the energy difference between the $n = 2$ and $n = 3$ levels is the energy of the photon. Refer to Figure 40.15 in the textbook.

EXECUTE: **(a)** We want $E_{1\text{-IDW}}$. Figure 40.15b, shows that $E_3 = 5.09E_{1\text{-IDW}}$ and $E_2 = 2.43E_{1\text{-IDW}}$. The energy difference between these levels is $E_3 - E_2 = (5.09 - 2.43)E_{1\text{-IDW}} = 2.50 \times 10^{-19}$ J, which gives $E_{1\text{-IDW}} = 9.40 \times 10^{-20}$ J.

(b) We want U_0. $U_0 = 6E_{1\text{-IDW}} = 6(9.40 \times 10^{-20} \text{ J}) = 5.64 \times 10^{-19}$ J.

(c) We want the width L of the well. Solve $E_{1\text{-IDW}} = n^2h^2/8mL^2$ for L and use the result from part (b) with $n = 1$. This gives $L = 0.800$ nm.

EVALUATE: The ground state energy of this well is $(0.625)E_{1\text{-IDW}} = 5.88 \times 10^{-20}$ J.

VP40.6.3. **IDENTIFY:** We are dealing with an electron in a finite potential well.

SET UP: $U_0 = 6E_{1\text{-IDW}}$, $E = hc/\lambda$. Use Figure 40.15 in the textbook.

EXECUTE: **(a)** We want the initial and final energy levels. In this case, there are only three possible transitions: $3 \rightarrow 2$, $2 \rightarrow 1$, and $3 \rightarrow 1$. The greatest energy difference is from the $3 \rightarrow 1$ transition which emits a photon of the shortest wavelength. So the initial state is $n = 3$ and the final state is $n = 1$.

(b) Figure 40.15b in the textbook gives $E_3 = 5.09E_{1\text{-IDW}}$ and $E_1 = 0.625E_{1\text{-IDW}}$. Therefore the energy of the photon is $E_3 - E_1 = (5.09 - 0625)E_{1\text{-IDW}} = 4.465E_{1\text{-IDW}}$. Using $E = hc/\lambda$ with the 355 nm wavelength and solving for $E_{1\text{-IDW}}$ gives $E_{1\text{-IDW}} = hc/[(4.465)(355 \text{ nm})]$.

$3 \rightarrow 2$ transition: The energy difference is $(5.09 - 2.43)E_{1\text{-IDW}}$, and this is the photon energy.

$$\frac{hc}{\lambda_{3 \rightarrow 2}} = 2.66E_{1\text{-IDW}} = (2.66)\frac{hc}{(4.465)(355 \text{ nm})}. \quad \lambda_{3 \rightarrow 2} = 596 \text{ nm}.$$

$2 \rightarrow 1$ transition: The energy difference is $(2.43 - 0.625)E_{1\text{-IDW}}$, and this is the photon energy.

$$\frac{hc}{\lambda_{2 \rightarrow 1}} = 1.805E_{1\text{-IDW}} = (1.805)\frac{hc}{(4.465)(355 \text{ nm})}. \quad \lambda_{2 \rightarrow 1} = 878 \text{ nm}.$$

EVALUATE: As the energy difference between the levels gets larger, the photon wavelength gets shorter, which is physically reasonable.

VP40.6.4. **IDENTIFY:** We are dealing with an electron in a finite potential well.

SET UP: $E_{1\text{-IDW}} = \dfrac{\pi^2 \hbar^2}{2mL^2}$.

EXECUTE: **(a)** We want the ground level energy for an infinite well. Using $L = 0.400$ nm and the equation for $E_{1\text{-IDW}}$ gives 2.35 eV.

(b) We want the energy of the bound state if $U_0 = 0.015$ eV. Because $U_0 \ll E_{1\text{-IDW}}$, $E = 0.68U_0$. Therefore $E = (0.68)(0.015 \text{ eV}) = 0.010$ eV.

EVALUATE: The energy of the bound state is much less than the ground level for an infinite well of the same length.

VP40.7.1. IDENTIFY: This problem is about an electron tunneling through a potential barrier. We want the probability that the electron will tunnel through barriers if different thicknesses if its energy is 3.75 eV and the barrier height is 6.10 eV.

SET UP: The probability T of tunneling us $T = Ge^{-2\kappa L}$, where

$$G = 16\frac{E}{U_0}\left(1 - \frac{E}{U_0}\right) \text{ and } \kappa = \sqrt{\frac{2m(U_0 - E)}{\hbar}}.$$

EXECUTE: (a) $L = 0.750$ nm. First calculate G and κ and then use them to find T.

$$G = 16\left(\frac{3.75 \text{ eV}}{6.10 \text{ eV}}\right)\left(1 - \frac{3.75 \text{ eV}}{6.10 \text{ eV}}\right) = 3.7893, \quad \kappa = \frac{\sqrt{2m(6.10 - 3.75) \text{ eV}}}{\hbar} = 7.85005 \times 10^9 \text{ m}^{-1}$$

Using these values gives $T = (3.789)e^{-11.763} = 2.92 \times 10^{-5}$.

(b) $L = 0.500$ nm. We get $T = (3.789)e^{-7.85005} = 1.48 \times 10^{-3}$.

EVALUATE: It is about 50 times more likely that the electron will tunnel through the narrower barrier than through the wider one. But probabilities are very low.

VP40.7.2. IDENTIFY: This problem is about an electron tunneling through a potential barrier.

SET UP: The probability T of tunneling us $T = Ge^{-2\kappa L}$, where

$$G = 16\frac{E}{U_0}\left(1 - \frac{E}{U_0}\right) \text{ and } \kappa = \sqrt{\frac{2m(U_0 - E)}{\hbar}}.$$

EXECUTE: (a) We want T when $L = 0.800$ nm. First calculate G and κ and then use them to find T.

$$G = 16\left(\frac{3.50 \text{ eV}}{4.00 \text{ eV}}\right)\left(1 - \frac{3.50 \text{ eV}}{4.00 \text{ eV}}\right) = 1.7500, \quad \kappa = \frac{\sqrt{2m(4.00 - 3.50) \text{ eV}}}{\hbar} = 3.62096 \times 10^9 \text{ m}^{-1}.$$

Using these values gives $T = (1.7500)e^{-5.79354} = 5.33 \times 10^{-3}$.

(b) We want L, so that the probability of tunneling is twice as great as in part (a). Solve $T = Ge^{-2\kappa L}$ for L. G and κ are the same as in part (a). Taking natural logarithms gives $L = -(1/2\kappa) \ln(T/G)$. Using G and κ with T twice what we found in part (a), we get $L = 0.704$ nm.

EVALUATE: Decreasing the width of the barrier from 0.800 nm to 0.704 nm doubled the probability of tunneling.

VP40.7.3. IDENTIFY: This problem is about an electron tunneling through a potential barrier. We want to find the probability of tunneling for different energies of the electron with $U_0 = 5.00$ eV and $L = 0.900$ nm.

SET UP: The probability T of tunneling us $T = Ge^{-2\kappa L}$, where

$$G = 16\frac{E}{U_0}\left(1 - \frac{E}{U_0}\right) \text{ and } \kappa = \frac{\sqrt{2m(U_0 - E)}}{\hbar}.$$

EXECUTE: (a) $E = 4.00$ eV. We want T when $L = 0.800$ nm. First calculate G and κ and then use them to find T.

$$G = 16\left(\frac{3.50 \text{ eV}}{4.00 \text{ eV}}\right)\left(1 - \frac{3.50 \text{ eV}}{4.00 \text{ eV}}\right) = 1.7500, \quad \kappa = \frac{\sqrt{2m(4.00 - 3.50) \text{ eV}}}{\hbar} = 3.62096 \times 10^9 \text{ m}^{-1}$$

Using these values gives $T = (1.7500)e^{-5.79354} = 5.33 \times 10^{-3}$.

(b) $E = 4.30$ eV. $G = 1.9264$, $\kappa = 4.2844 \times 10^9$ m^{-1}, $T = (1.9264)e^{-7.71188} = 8.62 \times 10^{-4}$.

(c) $E = 4.60$ eV. $G = 1.1776$, $\kappa = 3.23868 \times 10^9$ m^{-1}, $T = (1.1776)e^{-5.82963} = 3.46 \times 10^{-3}$.

EVALUATE: Our results show that as E gets closer to the height of the potential barrier, the probability of tunneling increases, which is physically reasonable.

VP40.7.4. **IDENTIFY:** This problem is about an electron tunneling through a potential barrier. We want to find the width L of the barrier if there is a 1/417 probability that the electron will tunnel through the barrier if the electron's energy is 3.00 eV and the barrier height is 5.00 eV.

SET UP: The probability T of tunneling us $T = Ge^{-2\kappa L}$, where

$$G = 16\frac{E}{U_0}\left(1 - \frac{E}{U_0}\right) \text{ and } \kappa = \frac{\sqrt{2m(U_0 - E)}}{\hbar}.$$

EXECUTE: Solve $T = Ge^{-2\kappa L}$ for L. Taking natural logarithms gives $L = -(1/2\kappa)\ln(T/G)$.

$$G = 16\left(\frac{3.00 \text{ eV}}{5.00 \text{ eV}}\right)\left(1 - \frac{3.00 \text{ eV}}{5.00 \text{ eV}}\right) = 3.8400, \ \kappa = \frac{\sqrt{2m(5.00 - 3.00) \text{ eV}}}{\hbar} = 7.24192 \times 10^9 \text{ m}^{-1}.$$

Using $T = 1/417$ and the G and κ we just calculated gives $L = 0.509$ nm.
EVALUATE: The value of L is comparable to atomic dimensions.

40.1. **IDENTIFY:** Using the momentum of the free electron, we can calculate k and ω and use these to express its wave function.

SET UP: $\Psi(x, t) = Ae^{ikx}e^{-i\omega t}$, $k = p/\hbar$, and $\omega = \hbar k^2/2m$.

EXECUTE: $k = \frac{p}{\hbar} = -\frac{4.50 \times 10^{-24} \text{ kg} \cdot \text{m/s}}{1.055 \times 10^{-34} \text{ J} \cdot \text{s}} = -4.27 \times 10^{10} \text{ m}^{-1}.$

$$\omega = \frac{\hbar k^2}{2m} = \frac{(1.055 \times 10^{-34} \text{ J} \cdot \text{s})(4.27 \times 10^{10} \text{ m}^{-1})^2}{2(9.108 \times 10^{-31} \text{ kg})} = 1.05 \times 10^{17} \text{ s}^{-1}.$$

$$\Psi(x, t) = Ae^{-i(4.27 \times 10^{10} \text{ m}^{-1})x}e^{-i(1.05 \times 10^{17} \text{ s}^{-1})t}.$$

EVALUATE: The wave function depends on position and time.

40.3. **IDENTIFY:** Use the wave function from Example 40.1.

SET UP: $|\Psi(x, t)|^2 = 2|A|^2\{1 + \cos[(k_2 - k_1)x - (\omega_2 - \omega_1)t]\}$. $k_2 = 3k_1 = 3k$. $\omega = \frac{\hbar k^2}{2m}$, so

$\omega_2 = 9\omega_1 = 9\omega$. $|\Psi(x, t)|^2 = 2|A|^2[1 + \cos(2kx - 8\omega t)]$.

EXECUTE: **(a)** At $t = 2\pi/\omega$, $|\Psi(x, t)|^2 = 2|A|^2[1 + \cos(2kx - 16\pi)]$. $|\Psi(x, t)|^2$ is maximum for $\cos(2kx - 16\pi) = 1$. This happens for $2kx - 16\pi = 0, 2\pi, \ldots$. Smallest positive x where $|\Psi(x, t)|^2$ is a maximum is $x = \frac{8\pi}{k}$.

(b) From the result of part (a), $v_{\text{av}} = \frac{8\pi/k}{2\pi/\omega} = \frac{4\omega}{k}$. $v_{\text{av}} = \frac{\omega_2 - \omega_1}{k_2 - k_1} = \frac{8\omega}{2k} = \frac{4\omega}{k}$.

EVALUATE: The two expressions agree.

40.7. **IDENTIFY:** We are dealing with a particle in a box.

SET UP: $E = hc/\lambda$, $E_n = n^2\frac{\pi^2\hbar^2}{2mL^2}$. We want the wavelength.

EXECUTE: The longest wavelength (lowest energy) photon is from a transition between the $n = 1$ to $n = 2$ states. The energy of this photon is equal to the energy difference between these states.

$$\Delta E_1 = \frac{\pi^2\hbar^2}{2mL^2}(2^2 - 1^2) = \frac{hc}{\lambda_1}.$$

The next longest wavelength photon is from a transition from the $n = 1$ to the $n = 3$ state.

$$\Delta E_2 = \frac{\pi^2 \hbar^2}{2mL^2}\left(3^2 - 1^2\right) = \frac{hc}{\lambda_2}.$$

Dividing these two equations and solving for λ_2 gives

$$\frac{4-1}{9-1} = \frac{\lambda_2}{\lambda_1} \cdot \lambda_2 = \frac{3}{8}\lambda_1 = \frac{3}{8}(420 \text{ nm}) = 158 \text{ nm}.$$

EVALUATE: The next longest photon (starting from the ground state) would be between the $n = 1$ and $n = 4$ states.

40.9. **IDENTIFY and SET UP:** The energy levels for a particle in a box are given by $E_n = \frac{n^2 h^2}{8mL^2}$.

EXECUTE: **(a)** The lowest level is for $n = 1$, and $E_1 = \frac{(1)(6.626\times10^{-34} \text{ J}\cdot\text{s})^2}{8(0.20 \text{ kg})(1.3 \text{ m})^2} = 1.6\times10^{-67} \text{ J}.$

(b) $E = \frac{1}{2}mv^2$, so $v = \sqrt{\frac{2E}{m}} = \sqrt{\frac{2(1.2\times10^{-67} \text{ J})}{0.20 \text{ kg}}} = 1.3\times10^{-33}$ m/s. If the ball has this speed the time it would take it to travel from one side of the table to the other is

$$t = \frac{1.3 \text{ m}}{1.3\times10^{-33} \text{ m/s}} = 1.0\times10^{33} \text{ s}.$$

(c) $E_1 = \frac{h^2}{8mL^2}$, $E_2 = 4E_1$, so $\Delta E = E_2 - E_1 = 3E_1 = 3(1.6\times10^{-67} \text{ J}) = 4.9\times10^{-67} \text{ J}.$

EVALUATE: **(d)** No, quantum mechanical effects are not important for the game of billiards. The discrete, quantized nature of the energy levels is completely unobservable.

40.11. **IDENTIFY:** An electron in the lowest energy state in this box must have the same energy as it would in the ground state of hydrogen.

SET UP: The energy of the nth level of an electron in a box is $E_n = \frac{nh^2}{8mL^2}$.

EXECUTE: An electron in the ground state of hydrogen has an energy of -13.6 eV, so find the width corresponding to an energy of $E_1 = 13.6$ eV. Solving for L gives

$$L = \frac{h}{\sqrt{8mE_1}} = \frac{(6.626\times10^{-34} \text{ J}\cdot\text{s})}{\sqrt{8(9.11\times10^{-31} \text{ kg})(13.6 \text{ eV})(1.602\times10^{-19} \text{ J/eV})}} = 1.66\times10^{-10} \text{ m}.$$

EVALUATE: This width is of the same order of magnitude as the diameter of a Bohr atom with the electron in the K shell.

40.17. **IDENTIFY:** We are dealing with a particle in a box.

SET UP: $E_n = n^2 \frac{\pi^2 \hbar^2}{2mL^2}$, $m_A = 9m_B$, $L_B = 2L_A$, $E_A = E_B$. We want the lowest possible quantum numbers n_A and n_B of the two states.

EXECUTE: Equate the energies and determine the lowest possible values of n_A and n_B.

$$\frac{n_A^2 \pi^2 \hbar^2}{2m_A^2 L_A^2} = \frac{n_B^2 \pi^2 \hbar^2}{2m_B^2 L_B^2} \cdot \frac{n_A^2}{(9m_B)(L_B/2)^2} = \frac{n_B^2}{m_B L_B^2} \cdot 4n_A^2 = 9n_B^2.$$

If $n_B = 2$, then $n_A = 3$, so the lowest possible values of the quantum numbers are $n_A = 3$, $n_B = 2$.

EVALUATE: Other states exist, such as $n_A = 6$, $n_B = 4$, but these are the lowest ones.

40.21. IDENTIFY: We are dealing with a particle in a box.

SET UP: $-\dfrac{\hbar^2}{2m}\dfrac{d^2\psi}{dx} = E\psi.$

EXECUTE: **(a)** $\psi_n = A\cos k_n x.$ Take the second derivative and use it to find E_n as follows.

$$-\frac{\hbar^2}{2m}\left(-Ak_n^2\cos k_n x\right) = EA\cos k_n x. \quad E_n = \frac{\hbar^2}{2m}k_n^2.$$

The wave function must be zero at $x = \pm L/2$. This gives
$A\cos\left[k_n(\pm L/2)\right] = 0.\, k_n(L/2) = \pi/2, 3\pi/2, 5\pi/2, \cdots .\, k_n = n\pi/L, n = 1, 3, 5, \cdots$

(b) $\psi_n = A\sin k_n x.$ Follow the same procedure as in part (a) to obtain $k_n = n\pi/L, n = 2, 4, 6, \ldots$
(c) We want the allowed energies.

$\psi_n = A\sin k_n x$: Combine the results for E_n and k_n from part (b) to obtain $E_n.$

$$E_n = \frac{\hbar^2}{2m}\left(\frac{n\pi}{L}\right)^2 = \frac{\hbar^2\pi^2}{2mL^2}n^2, n = 2, 4, 6, \cdots$$

$\psi_n = A\cos k_n x$: Follow the same procedure using the results from part (a), which gives

$$E_n = \frac{\hbar^2\pi^2}{2mL^2}n^2, n = 1, 3, 5, \cdots$$

EVALUATE: **(d)** The set of energies found here is the *same* as those in Eq. (40.31). This result occurs because the physical system (i.e., the box) does not "know" where we placed the origin of coordinates, so its behavior should be the same in either case.

40.25. IDENTIFY: The energy of the photon is the energy given to the electron.

SET UP: Since $U_0 = 6E_{1\text{-IDW}}$ we can use the result $E_1 = 0.625E_{1\text{-IDW}}$ from Section 40.4. When the electron is outside the well it has potential energy U_0, so the minimum energy that must be given to the electron is $U_0 - E_1 = 5.375E_{1\text{-IDW}}.$

EXECUTE: The maximum wavelength of the photon would be

$$\lambda = \frac{hc}{U_0 - E_1} = \frac{hc}{(5.375)(h^2/8mL^2)} = \frac{8mL^2c}{(5.375)h} = \frac{8(9.11\times10^{-31}\text{ kg})(1.50\times10^{-9}\text{ m})^2(3.00\times10^8\text{ m/s})}{(5.375)(6.63\times10^{-34}\text{ J}\cdot\text{s})}$$

$$= 1.38\times10^{-6}\text{ m}.$$

EVALUATE: This photon is in the infrared. The wavelength of the photon decreases when the width of the well decreases.

40.27. IDENTIFY: Find the transition energy ΔE and set it equal to the energy of the absorbed photon. Use $E = hc/\lambda$, to find the wavelength of the photon.

SET UP: $U_0 = 6E_{1\text{-IDW}}$, as in Figure 40.15 in the textbook, so $E_1 = 0.625E_{1\text{-IDW}}$ and

$E_3 = 5.09E_{1\text{-IDW}}$ with $E_{1\text{-IDW}} = \dfrac{\pi^2\hbar^2}{2mL^2}$. In this problem the particle bound in the well is a proton, so

$m = 1.673\times10^{-27}$ kg.

EXECUTE: $E_{1\text{-IDW}} = \dfrac{\pi^2 \hbar^2}{2mL^2} = \dfrac{\pi^2 (1.055 \times 10^{-34} \text{ J} \cdot \text{s})^2}{2(1.673 \times 10^{-27} \text{ kg})(4.0 \times 10^{-15} \text{ m})^2} = 2.052 \times 10^{-12}$ J. The transition

energy is $\Delta E = E_3 - E_1 = (5.09 - 0.625)E_{1\text{-IDW}} = 4.465 E_{1\text{-IDW}}$.

$\Delta E = 4.465(2.052 \times 10^{-12} \text{ J}) = 9.162 \times 10^{-12}$ J.

The wavelength of the photon that is absorbed is related to the transition energy by $\Delta E = hc/\lambda$, so

$\lambda = \dfrac{hc}{\Delta E} = \dfrac{(6.626 \times 10^{-34} \text{ J} \cdot \text{s})(2.998 \times 10^8 \text{ m/s})}{9.162 \times 10^{-12} \text{ J}} = 2.2 \times 10^{-14}$ m $= 22$ fm.

EVALUATE: The wavelength of the photon is comparable to the size of the well.

40.29. **IDENTIFY and SET UP:** The probability is $T = Ge^{-2\kappa L}$, with $G = 16\dfrac{E}{U_0}\left(1 - \dfrac{E}{U_0}\right)$ and

$\kappa = \dfrac{\sqrt{2m(U_0 - E)}}{\hbar}$. $E = 32$ eV, $U_0 = 41$ eV, $L = 0.25 \times 10^{-9}$ m. Calculate T.

EXECUTE: **(a)** $G = 16\dfrac{E}{U_0}\left(1 - \dfrac{E}{U_0}\right) = 16\dfrac{32}{41}\left(1 - \dfrac{32}{41}\right) = 2.741$.

$\kappa = \dfrac{\sqrt{2m(U_0 - E)}}{\hbar}$.

$\kappa = \dfrac{\sqrt{2(9.109 \times 10^{-31} \text{ kg})(41 \text{ eV} - 32 \text{ eV})(1.602 \times 10^{-19} \text{ J/eV})}}{1.055 \times 10^{-34} \text{ J} \cdot \text{s}} = 1.536 \times 10^{10}$ m^{-1}.

$T = Ge^{-2\kappa L} = (2.741)e^{-2(1.536 \times 10^{10} \text{ m}^{-1})(0.25 \times 10^{-9} \text{ m})} = 2.741e^{-7.68} = 0.0013$.

(b) The only change is the mass m, which appears in κ.

$\kappa = \dfrac{\sqrt{2m(U_0 - E)}}{\hbar}$.

$\kappa = \dfrac{\sqrt{2(1.673 \times 10^{-27} \text{ kg})(41 \text{ eV} - 32 \text{ eV})(1.602 \times 10^{-19} \text{ J/eV})}}{1.055 \times 10^{-34} \text{ J} \cdot \text{s}} = 6.584 \times 10^{11}$ m^{-1}.

Then $T = Ge^{-2\kappa L} = (2.741)e^{-2(6.584 \times 10^{11} \text{ m}^{-1})(0.25 \times 10^{-9} \text{ m})} = 2.741e^{-392.2} = 10^{-143}$.

EVALUATE: The more massive proton has a much smaller probability of tunneling than the electron does.

40.31. **IDENTIFY:** The tunneling probability is $T = 16\dfrac{E}{U_0}\left(1 - \dfrac{E}{U_0}\right)e^{-2L\sqrt{2m(U_0 - E)}/\hbar}$.

SET UP: $\dfrac{E}{U_0} = \dfrac{6.0 \text{ eV}}{11.0 \text{ eV}}$ and $E - U_0 = 5 \text{ eV} = 8.0 \times 10^{-19}$ J.

EXECUTE: **(a)** $L = 0.80 \times 10^{-9}$ m:

$T = 16\left(\dfrac{6.0 \text{ eV}}{11.0 \text{ eV}}\right)\left(1 - \dfrac{6.0 \text{ ev}}{11.0 \text{ eV}}\right)e^{-2(0.80 \times 10^{-9} \text{ m})\sqrt{2(9.11 \times 10^{-31} \text{ kg})(8.0 \times 10^{-19} \text{ J})}/1.055 \times 10^{-34} \text{ J} \cdot \text{s}} = 4.4 \times 10^{-8}$.

(b) $L = 0.40 \times 10^{-9}$ m: $T = 4.2 \times 10^{-4}$.

EVALUATE: The tunneling probability is less when the barrier is wider.

40.33. **IDENTIFY** and **SET UP:** The energy levels are given by $E_n = (n + \frac{1}{2})\hbar\omega$, where $\omega = \sqrt{\dfrac{k'}{m}}$.

EXECUTE: $\omega = \sqrt{\dfrac{k'}{m}} = \sqrt{\dfrac{110 \text{ N/m}}{0.250 \text{ kg}}} = 21.0$ rad/s.

The ground state energy is given by $E_n = (n + \frac{1}{2})\hbar\omega$, where $n = 0$.

$E_0 = \frac{1}{2}\hbar\omega = \frac{1}{2}(1.055 \times 10^{-34} \text{ J} \cdot \text{s})(21.0 \text{ rad/s}) = 1.11 \times 10^{-33} \text{ J}(1 \text{ eV}/1.602 \times 10^{-19} \text{ J}) = 6.93 \times 10^{-15}$ eV.

$E_n = (n + \frac{1}{2})\hbar\omega$, $E_{(n+1)} = (n + 1 + \frac{1}{2})\hbar\omega$.

The energy separation between these adjacent levels is

$\quad \Delta E = E_{n+1} - E_n = \hbar\omega = 2E_0 = 2(1.11 \times 10^{-33} \text{ J}) = 2.22 \times 10^{-33} \text{ J} = 1.39 \times 10^{-14}$ eV.

EVALUATE: These energies are extremely small; quantum effects are not important for this oscillator.

40.35. **IDENTIFY:** We can model the molecule as a harmonic oscillator. The energy of the photon is equal to the energy difference between the two levels of the oscillator.

SET UP: The energy of a photon is $E_\gamma = hf = hc/\lambda$, and the energy levels of a harmonic oscillator are given by $E_n = (n + \frac{1}{2})\hbar\sqrt{\dfrac{k'}{m}} = (n + \frac{1}{2})\hbar\omega$.

EXECUTE: **(a)** The photon's energy is $E_\gamma = \dfrac{hc}{\lambda} = \dfrac{(6.63 \times 10^{-34} \text{ J} \cdot \text{s})(3.00 \times 10^8 \text{ m/s})}{5.8 \times 10^{-6} \text{ m}} = 0.21$ eV.

(b) The transition energy is $\Delta E = E_{n+1} - E_n = \hbar\omega = \hbar\sqrt{\dfrac{k'}{m}}$, which gives $\dfrac{2\pi\hbar c}{\lambda} = \hbar\sqrt{\dfrac{k'}{m}}$. Solving for

k', we get $k' = \dfrac{4\pi^2 c^2 m}{\lambda^2} = \dfrac{4\pi^2 (3.00 \times 10^8 \text{ m/s})^2 (5.6 \times 10^{-26} \text{ kg})}{(5.8 \times 10^{-6} \text{ m})^2} = 5{,}900$ N/m.

EVALUATE: This would be a rather strong spring in the physics lab.

40.37. **IDENTIFY:** The photon energy equals the transition energy for the atom.

SET UP: According to the energy level equation $E_n = (n + \frac{1}{2})\hbar\omega$, the energy released during the transition between two adjacent levels is twice the ground state energy $E_3 - E_2 = \hbar\omega = 2E_0 = 11.2$ eV.

EXECUTE: For a photon of energy E,

$\quad E = hf \Rightarrow \lambda = \dfrac{c}{f} = \dfrac{hc}{E} = \dfrac{(6.63 \times 10^{-34} \text{ J} \cdot \text{s})(3.00 \times 10^8 \text{ m/s})}{(11.2 \text{ eV})(1.60 \times 10^{-19} \text{ J/eV})} = 111$ nm.

EVALUATE: This photon is in the ultraviolet.

40.41. **IDENTIFY:** We know the wave function of a particle in a box.

SET UP and EXECUTE: **(a)** $\Psi(x, t) = \dfrac{1}{\sqrt{2}}\psi_1(x)e^{-iE_1 t/\hbar} + \dfrac{1}{\sqrt{2}}\psi_3(x)e^{-iE_3 t/\hbar}$.

$\quad\quad \Psi^*(x, t) = \dfrac{1}{\sqrt{2}}\psi_1(x)e^{+iE_1 t/\hbar} + \dfrac{1}{\sqrt{2}}\psi_3(x)e^{+iE_3 t/\hbar}$.

$$|\Psi(x,t)|^2 = \tfrac{1}{2}[\psi_1^2 + \psi_3^2 + \psi_1\psi_3(e^{i(E_3-E_1)t/\hbar} + e^{-i(E_3-E_1)t/\hbar})] = \tfrac{1}{2}\left[\psi_1^2 + \psi_3^2 + 2\psi_1\psi_3 \cos\left(\frac{[E_3-E_1]t}{\hbar}\right)\right].$$

$$\psi_1 = \sqrt{\frac{2}{L}}\sin\left(\frac{\pi x}{L}\right), \psi_3 = \sqrt{\frac{2}{L}}\sin\left(\frac{3\pi x}{L}\right). E_3 = \frac{9\pi^2\hbar^2}{2mL^2} \text{ and } E_1 = \frac{\pi^2\hbar^2}{2mL^2}, \text{ so } E_3 - E_1 = \frac{4\pi^2\hbar^2}{mL^2}.$$

$$|\Psi(x,t)|^2 = \frac{1}{L}\left[\sin^2\left(\frac{\pi x}{L}\right) + \sin^2\left(\frac{3\pi x}{L}\right) + 2\sin\left(\frac{\pi x}{L}\right)\sin\left(\frac{3\pi x}{L}\right)\cos\left(\frac{4\pi^2\hbar t}{mL^2}\right)\right]. \text{ At } x = L/2,$$

$$\sin\left(\frac{\pi x}{L}\right) = \sin\left(\frac{\pi}{2}\right) = 1. \sin\left(\frac{3\pi x}{L}\right) = \sin\left(\frac{3\pi}{2}\right) = -1. |\Psi(x,t)|^2 = \frac{2}{L}\left[1 - \cos\left(\frac{4\pi^2\hbar t}{mL^2}\right)\right].$$

(b) $\omega_{osc} = \dfrac{E_3 - E_1}{\hbar} = \dfrac{4\pi^2\hbar}{mL^2}.$

EVALUATE: Note that $\Delta E = \hbar\omega$.

40.47. **IDENTIFY** and **SET UP:** The normalized wave function for the $n = 2$ first excited level is

$\psi_2 = \sqrt{\dfrac{2}{L}}\sin\left(\dfrac{2\pi x}{L}\right). P = |\psi(x)|^2\, dx$ is the probability that the particle will be found in the interval x to $x + dx$.

EXECUTE: (a) $x = L/4.$

$$\psi(x) = \sqrt{\frac{2}{L}}\sin\left(\left(\frac{2\pi}{L}\right)\left(\frac{L}{4}\right)\right) = \sqrt{\frac{2}{L}}\sin\left(\frac{\pi}{2}\right) = \sqrt{\frac{2}{L}}.$$
$$P = (2/L)dx.$$

(b) $x = L/2.$

$$\psi(x) = \sqrt{\frac{2}{L}}\sin\left(\left(\frac{2\pi}{L}\right)\left(\frac{L}{2}\right)\right) = \sqrt{\frac{2}{L}}\sin(\pi) = 0.$$
$$P = 0.$$

(c) $x = 3L/4.$

$$\psi(x) = \sqrt{\frac{2}{L}}\sin\left(\left(\frac{2\pi}{L}\right)\left(\frac{3L}{4}\right)\right) = \sqrt{\frac{2}{L}}\sin\left(\frac{3\pi}{2}\right) = -\sqrt{\frac{2}{L}}.$$
$$P = (2/L)dx.$$

EVALUATE: Our results are consistent with the $n = 2$ part of Figure 40.12 in the textbook. $|\psi|^2$ is zero at the center of the box and is symmetric about this point.

40.49. **IDENTIFY:** The probability of the particle being between x_1 and x_2 is $\int_{x_1}^{x_2}|\psi|^2\, dx$, where ψ is the normalized wave function for the particle.

(a) SET UP: The normalized wave function for the ground state is $\psi_1 = \sqrt{\dfrac{2}{L}}\sin\left(\dfrac{\pi x}{L}\right).$

EXECUTE: The probability P of the particle being between $x = L/4$ and $x = 3L/4$ is

$P = \int_{L/4}^{3L/4}|\psi_1|^2\, dx = \dfrac{2}{L}\int_{L/4}^{3L/4}\sin^2\left(\dfrac{\pi x}{L}\right)dx.$ Let $y = \pi x/L; dx = (L/\pi)dy$ and the integration limits become $\pi/4$ and $3\pi/4.$

$$P = \frac{2}{L}\left(\frac{L}{\pi}\right)\int_{\pi/4}^{3\pi/4} \sin^2 y\, dy = \frac{2}{\pi}\left[\frac{1}{2}y - \frac{1}{4}\sin 2y\right]_{\pi/4}^{3\pi/4}.$$

$$P = \frac{2}{\pi}\left[\frac{3\pi}{8} - \frac{\pi}{8} - \frac{1}{4}\sin\left(\frac{3\pi}{2}\right) + \frac{1}{4}\sin\left(\frac{\pi}{2}\right)\right].$$

$$P = \frac{2}{\pi}\left(\frac{\pi}{4} - \frac{1}{4}(-1) + \frac{1}{4}(1)\right) = \frac{1}{2} + \frac{1}{\pi} = 0.818. \quad \text{(Note: The integral formula } \int \sin^2 y\, dy = \frac{1}{2}y - \frac{1}{4}\sin 2y$$

was used.)

(b) SET UP: The normalized wave function for the first excited state is $\psi_2 = \sqrt{\dfrac{2}{L}}\sin\left(\dfrac{2\pi x}{L}\right)$.

EXECUTE: $P = \int_{L/4}^{3L/4} |\psi_2|^2 dx = \dfrac{2}{L}\int_{L/4}^{3L/4}\sin^2\left(\dfrac{2\pi x}{L}\right) dx.$ Let $y = 2\pi x/L$; $dx = (L/2\pi)\,dy$ and the integration limits become $\pi/2$ and $3\pi/2$.

$$P = \frac{2}{L}\left(\frac{L}{2\pi}\right)\int_{\pi/2}^{3\pi/2} \sin^2 y\, dy = \frac{1}{\pi}\left[\frac{1}{2}y - \frac{1}{4}\sin 2y\right]_{\pi/2}^{3\pi/2} = \frac{1}{\pi}\left(\frac{3\pi}{4} - \frac{\pi}{4}\right) = 0.500.$$

EVALUATE: **(c)** These results are consistent with Figure 40.11b in the textbook. That figure shows that $|\psi|^2$ is more concentrated near the center of the box for the ground state than for the first excited state; this is consistent with the answer to part (a) being larger than the answer to part (b). Also, this figure shows that for the first excited state half the area under $|\psi|^2$ curve lies between $L/4$ and $3L/4$, consistent with our answer to part (b).

40.53. IDENTIFY: This problem is about a quantum mechanical harmonic oscillator.

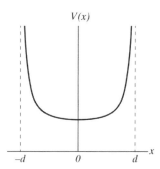

Figure 40.53

SET UP and EXECUTE: **(a)** For the sketch, see Figure 40.53.
(b) First combine the terms in the given equation for $V(x)$ to obtain the following:

$$V(x) = \frac{2kq^2}{d}\frac{1}{1-(x/d)^2}, \text{ where } k=1/4\pi\epsilon_0 .$$

Expand the right-hand fraction using $(1 + z)^n \approx 1 + nz$, where $z \ll 1$, $n = -1$, $z = -(x/d)^2$.

$$V(x) = \frac{2kq^2}{d}\left[1 + \frac{x^2}{d^2}\right].$$

Using $q = 6e$ and putting in k gives

$$V(x) = \frac{18e^2}{\pi\epsilon_0 d} + \frac{18e^2}{\pi\epsilon_0 d^3}x^2.$$

(c) We want the spring constant. Using the second term for $V(x)$, we see that

$$V = \frac{1}{2}k'x^2 = \frac{18e^2}{\pi\epsilon_0 d^3}x^2$$

$$k = \frac{36e^2}{\pi\epsilon_0 d^3} = 265 \text{ N/m}.$$

(d) The classical ground state is $x = 0$, so the energy at this state is

$$V = \frac{18e^2}{\pi\epsilon_0 d} = 207 \text{ eV}.$$

(e) We want the energy of the lowest energy photon. Use the quantum energy states.

$$E_0 = \frac{1}{2}\hbar\sqrt{\frac{k'}{m}} \text{ and } E_n = \left(n+\frac{1}{2}\right)\hbar\sqrt{\frac{k'}{m}}, \text{ so } E_n = 2E_0\left(n+\frac{1}{2}\right).$$

The lowest energy photon is due to a transition from the $n = 1$ state to the $n = 0$ state, so

$$\Delta E = 2E_0\left[\left(1+\frac{1}{2}\right)-\left(0+\frac{1}{2}\right)\right] = 2E_0 = 2(0.0379 \text{ eV}) = 0.0758 \text{ eV}.$$

(f) Use $E = hc/\lambda$ with $E = 0.0758$ eV, giving $\lambda = 16.4$ μm.
EVALUATE: There is obviously a big difference between a quantum oscillator and a classical oscillator.

40.55. **IDENTIFY** and **SET UP:** Calculate the angular frequency ω of the pendulum and apply $E_n = (n+\frac{1}{2})\hbar\omega$ for the energy levels.

EXECUTE: $\omega = \frac{2\pi}{T} = \frac{2\pi}{0.500 \text{ s}} = 4\pi \text{ s}^{-1}$.

The ground-state energy is $E_0 = \frac{1}{2}\hbar\omega = \frac{1}{2}(1.055\times10^{-34} \text{ J·s})(4\pi \text{ s}^{-1}) = 6.63\times10^{-34}$ J.

$$E_0 = 6.63\times10^{-34} \text{ J}(1 \text{ eV}/1.602\times10^{-19} \text{ J}) = 4.14\times10^{-15} \text{ eV}.$$

$$E_n = (n+\tfrac{1}{2})\hbar\omega.$$

$$E_{n+1} = (n+1+\tfrac{1}{2})\hbar\omega.$$

The energy difference between the adjacent energy levels is

$$\Delta E = E_{n+1} - E_n = \hbar\omega = 2E_0 = 1.33\times10^{-33} \text{ J} = 8.30\times10^{-15} \text{ eV}.$$

EVALUATE: These energies are much too small to detect. Quantum effects are not important for ordinary size objects.

40.59. **IDENTIFY** and **SET UP:** The energy levels for an infinite potential well are $E_n = \frac{n^2h^2}{8mL^2} = E_1n^2$. The energy of the absorbed photon is equal to the energy difference between the levels. The energy of a photon is $E = hf = hc/\lambda$, so $\Delta E = hf = E_{n+1} - E_n$.

EXECUTE: **(a)** For the first transition, we have $hf_1 = E_1(n^2 - 1^2)$, and for the second transition we have

$hf_2 = E_1[(n+1)^2 - 1^2]$. Taking the ratio of these two equations gives

$$\frac{hf_2}{hf_1} = \frac{16.9}{9.0} = \frac{(n+1)^2-1}{n^2-1} = \frac{n^2+2n+1-1}{n^2-1} = \frac{n^2+2n}{n^2-1}.$$

Rearranging and collecting terms gives the quadratic equation $n^2\left(\dfrac{16.9}{9.0}-1\right)-2n-\dfrac{16.9}{9.0}=0.$ Using the quadratic formula and taking the positive root gives $n = 3.0$, so $n = 3$. Therefore the transitions are from the $n = 3$ and $n = 4$ levels to the $n = 1$ level.

(b) Using the $3 \to 1$ transition with $f_1 = 9.0 \times 10^{14}$ Hz, we have

$$hf_1 = (h^2/8mL^2)(3^2 - 1^2) = h^2/mL^2.$$

$$L = \sqrt{\dfrac{h}{f_1 m}} = \sqrt{\dfrac{6.626 \times 10^{-34}\ \text{J} \cdot \text{s}}{(9.0 \times 10^{14}\ \text{Hz})(9.109 \times 10^{-31}\ \text{kg})}} = 9.0 \times 10^{-10}\ \text{m} = 0.90\ \text{nm}.$$

(c) The longest wavelength is for the smallest energy, and that would be for a transition between $n = 1$ and $n = 2$ levels. Comparing the $1 \to 3$ transition and the $1 \to 2$ transition, we have

$$\dfrac{hf_{1\to2}}{hf_{1\to3}} = \dfrac{E_1(2^2 - 1^2)}{E_1(3^2 - 1^2)} \quad \to \quad f_{1\to2} = \dfrac{3}{8}f_{1\to3} = \dfrac{3}{8}(9.0 \times 10^{14}\ \text{Hz}).$$

$$\lambda = c/f = \dfrac{3.00 \times 10^8\ \text{m/s}}{\dfrac{3}{8}(9.0 \times 10^{14}\ \text{Hz})} = 890\ \text{nm}.$$

EVALUATE: This wavelength is too long to be visible light. The wavelength of the 9.0×10^{14} Hz photon is 333 nm, which is too short to be visible, as is the 16.9×10^{14} Hz photon. So none of these photons will be visible.

40.61. **IDENTIFY** and **SET UP:** The transmission coefficient T is equal to 1 when the width L of the barrier is $L = \frac{1}{2}\lambda,\ \lambda,\ \frac{3}{2}\lambda,\ 2\lambda,\ \ldots = n\lambda/2$, where $n = 1, 2, 3, \ldots$, and where λ is the de Broglie wavelength of the electron, given by $\lambda = h/p$. The total energy of the electron is $E = U + K$, and $K = p^2/2m$.

EXECUTE: From the condition on λ, we have $\lambda_n = 2L/n$. Therefore $\lambda = h/p = 2L/n$, which gives $p = nh/2L$. The kinetic energy is $K = p^2/2m$, so $K_n = \dfrac{p^2}{2m} = \dfrac{\left(\dfrac{nh}{2L}\right)^2}{2m} = n^2\left(\dfrac{h^2}{8mL^2}\right) = n^2 K_1.$ The three lowest values of K are for $n = 1, 2,$ and 3.
$K_1 = h^2/8mL^2 = (6.626 \times 10^{-34}\ \text{J} \cdot \text{s})^2/[8(9.11 \times 10^{-31}\ \text{kg})(1.8 \times 10^{-10}\ \text{m})^2] = 1.86 \times 10^{-18}\ \text{J} = 11.6\ \text{eV}.$
The total energy is $E = K + U$, so $E_1 = K_1 + U = 11.6\ \text{eV} + 10\ \text{eV} = 22\ \text{eV}.$
For the $n = 2$ state, we have
$\quad K_2 = 2^2 K_1 = 4(11.6\ \text{eV}) = 46.4\ \text{eV}$, so $E_2 = 46.4\ \text{eV} + 10\ \text{eV} = 56\ \text{eV}.$
For the $n = 3$ state, we have
$K_3 = 3^2 K_1 = 9(11.6\ \text{eV}) = 104.4\ \text{eV}$, so $E_3 = 104.4\ \text{eV} + 10\ \text{eV} = 114\ \text{eV},$ which rounds to 110 eV.
EVALUATE: We cannot use Eq. (40.42) because T is not small.

40.63. **IDENTIFY:** This problem involves commutators and the Schrödinger equation.

SET UP: $[A, B]f = A(Bf) - B(Af)$. The following operators are defined:

$$a_+ = \dfrac{1}{\sqrt{2m\hbar\omega}}\left(-\hbar\dfrac{d}{dx} + m\omega x\right) \quad \text{and} \quad a_- = \dfrac{1}{\sqrt{2m\hbar\omega}}\left(+\hbar\dfrac{d}{dx} + m\omega x\right).$$

EXECUTE: **(a)** We want $[a_-, a_+]$.

$$[a_-, a_+]f = a_-(a_+f) - a_+(a_-f) = \frac{1}{\sqrt{2m\hbar\omega}} \frac{1}{\sqrt{2m\hbar\omega}} \left(\hbar\frac{d}{dx} + m\omega x\right)\left[\left(-\hbar\frac{d}{dx} + m\omega x\right)f\right]$$

$$-\frac{1}{\sqrt{2m\hbar\omega}} \frac{1}{\sqrt{2m\hbar\omega}} \left(-\hbar\frac{d}{dx} + m\omega x\right)\left[\left(\hbar\frac{d}{dx} + m\omega x\right)f\right]$$

Carefully carrying out all the operations gives $[a_-, a_+]f = f$, so $[a_-, a_+] = 1$.

(b) We want $[a_+, a_-]\psi$, where ψ is a wave equation, which means that it is a solution to the Schrödinger equation. Follow the procedure in part (a). The solution is sketched out below.

$$a_+a_-\psi = \frac{1}{\sqrt{2m\hbar\omega}}\left(-\hbar\frac{d}{dx} + m\omega x\right)\left[\frac{1}{\sqrt{2m\hbar\omega}}\left(\hbar\frac{d}{dx} + m\omega x\right)\psi\right].$$

$$a_+a_-\psi = \frac{1}{\sqrt{2m\hbar\omega}}\left(-\hbar\frac{d}{dx} + m\omega x\right)\left(\hbar\frac{d\psi}{dx} + m\omega x\psi\right).$$

$$a_+a_-\psi = \frac{1}{\sqrt{2m\hbar\omega}}\left(-\hbar^2\frac{d^2\psi}{dx^2} - \hbar m\omega\frac{d(x\psi)}{dx} + m\omega x\frac{d\psi}{dx} + m^2\omega^2 x^2\psi\right).$$

$$a_+a_-\psi = \frac{1}{\sqrt{2m\hbar\omega}}\left(-\hbar^2\frac{d^2\psi}{dx^2} - \hbar m\omega\psi + m^2\omega^2 x^2\psi\right) = \frac{1}{\hbar\omega}\left(-\frac{\hbar^2}{2m}\psi'' + \frac{m\omega^2 x^2}{2}\psi\right) - \frac{1}{2}\psi.$$

(c) Use the results of part (b) and the Schrödinger equation given with this problem. Note that the term in parentheses in the answer to (b) is the Schrödinger equation. Solving for it gives

$$\hbar\omega\left(a_+a_-\psi + \frac{1}{2}\psi\right) = -\frac{\hbar^2}{2m}\psi'' + \frac{m\omega^2 x^2}{2}\psi$$

Therefore the operator H must be

$$H = \hbar\omega\left(a_+a_- + \frac{1}{2}\right).$$

(d) We want $[H, a_+]$. Using $[H, a_\pm]f = H(a_+f) - a_+(Hf)$ gives

$$[H, a_+]f = H(a_+f) - a_+\left[\hbar\omega\left(a_+a_- + \frac{1}{2}\right)\right]f = \hbar\omega\left[\left(\frac{1}{2} + a_+a_-\right)(a_+f) - a_+\frac{f}{2} - a_+(a_+a_-f)\right]$$

$$[H, a_+]f = \hbar\omega(a_+)(a_-a_+ - a_+a_-)f = \hbar\omega a_+f.$$

The same procedure leads to a similar result for $[H, a_\pm]$. Therefore the commutators are

$$[H, a_+] = \hbar\omega a_+ \quad \text{and} \quad [H, a] = \hbar\omega a_-.$$

(e) We want to relate E_{n+1} to E_n.

For the n state: $H\Psi_n = E_n\Psi_n$

For the $n+1$ state: $H\psi_{n+1} = E_{n+1}\psi_{n+1}$ and $\psi_{n+1} = a_+\psi_n$, so $H\psi_{n+1} = Ha_+\psi_n$

$$[H, a_+]\psi_n = Ha_+\psi_n - a_+H\psi_n \quad \text{gives} \quad Ha_+\psi_n = [H, a_+]\psi_n + a_+H\psi_n$$

$$H\psi_{n+1} = Ha_+\psi_n = ([H, a_+] + a_+H)\psi_n$$

Using the result from part (d), we can write this result as

$$H\psi_{n+1} = a_+H\psi_n + \hbar\omega a_+\psi_n$$

$H\psi_n = E_n\psi_n$ and $\psi_{n+1} = a_+\psi_n$ gives $H\psi_{n+1} = E_n\psi_{n+1} + \hbar\omega\psi_{n+1} = (E_n + \hbar\omega)\psi_{n+1}$

This result means that $E_{n+1} = E_n + \hbar\omega$.

(f) We want the ground state energy E_0.

Using $H\psi_0 = E_0\psi_0$ gives $\hbar\omega\left(\dfrac{1}{2} + a_+a_-\right)\psi_0 = E_0\psi_0$. Given that $a_-\psi_0 = 0$, we have

$\dfrac{1}{2}\hbar\omega\psi_0 = E_0\psi_0$, which tells us that $E_0 = \dfrac{1}{2}\hbar\omega$.

(g) We want the energy levels E_n. From (e) we know that the energy of each state is $\hbar\omega$ higher than the previous state, and from (f) we know that the energy of the lowest state is $\hbar\omega/2$. Therefore the energies are

$$E_0 = \hbar\omega\left(\dfrac{1}{2}\right), \; E_1 = \hbar\omega\left(1+\dfrac{1}{2}\right), \; E_2 = \hbar\omega\left(2+\dfrac{1}{2}\right), \; E_3 = \hbar\omega\left(3+\dfrac{1}{2}\right), \cdots. \; E_n = \hbar\omega\left(n+\dfrac{1}{2}\right).$$

EVALUATE: The use of the method developed here depends on selecting appropriate operators $a_\pm$.

41

STUDY GUIDE FOR ATOMIC STRUCTURE

Summary

In this chapter, we'll apply our knowledge of quantum mechanics and apply the Schrödinger equation to three-dimensional problems atoms in order to understand the structure at atoms. We'll learn how the quantization of angular momentum is a natural result of our investigation. In addition, we'll see how atoms are described in terms of their quantum numbers, and we'll learn that electrons have an intrinsic spin quantum number. We'll also learn how the Pauli exclusionary principle prevents two particles from occupying the same quantum mechanical state.

Objectives

After studying this chapter, you'll understand

- How to apply the Schrödinger equation to three-dimensional problems.

- How to describe the states of the hydrogen atom in terms of quantum numbers.

- How the Zeeman effect describes the orbital motion of atomic electrons in a magnetic field.

- That electrons have intrinsic spin angular momentum.

- How to analyze the structure of many-electron atoms.

- How X-rays emitted by atoms unveil the inner structure of atoms.

Concepts and Equations

Term	Description
The Hydrogen Atom	The Schrödinger equation predicts the same energy levels as the Bohr model: $$E_n = -\frac{1}{(4\pi\varepsilon_0)^2}\frac{m_r e^4}{2n^2\hbar^2} = -\frac{13.60\,\text{eV}}{n^2}.$$ It also gives the possible magnitudes of orbital angular momentum as $$L = \sqrt{l(l+1)}\;\hbar, \qquad l = 0, 1, 2, \ldots,\; n-1,$$ and the z component of the orbital angular momentum as $$L_z = m_l \hbar, \quad m_l = 0, \pm 1, \pm 2, \ldots, \pm l.$$
The Zeeman Effect	The interaction energy of an electron with magnetic quantum number m_l in a magnetic field along the $+z$-axis is given by $$U = -\mu_z B = m_l \frac{e\hbar}{2m} B = m_l \mu_B B, \qquad m_l = 0, \pm 1, \pm 2, \ldots, \pm l,$$ where υ_B is the Bohr magneton.
Electron Spin	Electrons have intrinsic spin angular momentum of magnitude S, given by $$S = \sqrt{\frac{1}{2}\left(\frac{1}{2}+1\right)}\;\hbar = \sqrt{\frac{3}{4}}\;\hbar.$$ The z-component of the spin angular momentum has values $$S_z = m_s \hbar, \qquad m_s = \pm\tfrac{1}{2}.$$
Electrons in Atoms	In a hydrogen atom, the quantum numbers (n, l, m_l, m_s) specify the quantum-mechanical state of the atom and have allowed values given by $$n \geq 1, \qquad 0 \leq l \leq n-1,$$ $$\lvert m_l \rvert \leq l, \qquad m_s = \pm\tfrac{1}{2}.$$
X-ray Spectra	Moseley's law states that the frequency of the K_α X-ray from a target with atomic number Z is given by $$f = (2.48 \times 10^{15}\,\text{Hz})(Z-1)^2.$$

Key Concept 1: If a particle is in a three-dimensional stationary state described by wave function $\psi(x,y,z)$, the probability of finding that particle in an infinitesimal volume dV centered on (x, y, z) is $\lvert \psi(x,y,z) \rvert^2 dV$. The probability of finding the particle somewhere in a finite volume equals the integral $\int \lvert \psi(x,y,z) \rvert^2 dV$ evaluated over that volume.

Key Concept 2: Because the potential-energy function for the hydrogen atom is spherically symmetric, its energy levels are degenerate. Stationary states with the same principal quantum number n have the same energy, even though they have different values of the orbital quantum number l and the orbital magnetic quantum number m_l.

Key Concept 3: For each value of the principal quantum number $n = 1, 2, 3, \ldots$ for the stationary states of the hydrogen atom, there are n possible values of the orbital quantum number l that range from 0 to $n-1$. For each value of l, there are $2l + 1$ possible values of the orbital magnetic quantum number m_l that range from $-l$ to l.

From Chapter 41 of Student's Study Guide to accompany *University Physics with Modern Physics, Volume 3,* Fifteenth Edition. Hugh D. Young and Roger A. Freedman. Copyright © 2020 by Pearson Education, Inc. All rights reserved.

Key Concept 4: If a particle is in a stationary state described by the wave function ψ, the probability of finding the particle at a radial coordinate between r and $r + dr$ is $P(r)dr$, where $P(r) = 4\pi r^2 |\psi|^2$. If the wave function depends on the angle as well as on the radial coordinate, you must first average $|\psi|^2$ over all angles.

Key Concept 5: An electron in an atom has a nonzero magnetic moment if it is in a state with a nonzero orbital angular momentum. If this atom is in a magnetic field $\vec{B}$, the electron energy depends on the electron's orbital magnetic quantum number m_l, which describes how the magnetic moment vector is oriented with respect to $\vec{B}$.

Key Concept 6: In addition to any orbital angular momentum, an electron has an intrinsic angular momentum $\vec{S}$ called spin. The component of $\vec{S}$ along a given axis can have only two possible values, $+\frac{1}{2}\hbar$ ("spin up") or $-\frac{1}{2}\hbar$ ("spin down"). An electron also has a spin magnetic moment $\vec{\mu}$ directed opposite to $\vec{S}$, so when it is placed in a magnetic field $\vec{B}$ the electron's energy depends on the component of $\vec{S}$ along the direction of $\vec{B}$.

Key Concept 7: An atomic electron with an orbital angular momentum around the nucleus experiences an effective magnetic field. This field interacts with the spin magnetic moment of the electron, so the electron energy depends on the relative orientation of its spin angular momentum and orbital angular momentum. This is called spin-orbit coupling.

Key Concept 8: When both magnetic and relativistic effects are included, the energy levels of the hydrogen atom [Eq. (41.41)] depend on both the principal quantum number $n = 1, 2, 3, \ldots$ and the quantum number j associated with the electron's total angular momentum (the vector sum of the orbital and spin angular momenta). If the electron has orbital quantum number l, the value of j is either $l + \frac{1}{2}$ or $\left| l - \frac{1}{2} \right|$. Transitions between states that involve emitting or absorbing a photon are allowed only if l changes by 1 in the transition.

Key Concept 9: If a many-electron atom has just one outer valence electron, the valence electron's energy levels are like those of a single electron atom but with the atomic number Z of the nucleus replaced by an effective atomic number Z_{eff}. The value of Z_{eff} is less than Z because the inner electrons partially "screen" the electric field of the nucleus.

Key Concept 10: For a many-electron atom with just one outer valence electron, if that electron is in a state of high orbital angular momentum it "sees" the effective atomic number Z_{eff} of the nucleus to be 1. In such a state the valence electron is almost always outside of the other $Z - 1$ electrons, which "screen" $(Z - 1)e$ of the nuclear charge $+Ze$.

Key Concept 11: If an electron is kicked out of the innermost, or K, shell of an atom of atomic number Z, then an electron from the next shell (L) drops down to the K shell to fill in the vacancy and emits an X-ray photon in the process. The frequency of this photon is proportional to $(Z - 1)^2$ (Moseley's law).

Conceptual Questions

1: Atoms without the Pauli exclusion principle

What would the electron configuration of the ground state of calcium be if the Pauli exclusion principle did not hold?

IDENTIFY, SET UP, AND EXECUTE Without the exclusion principle, all electrons could occupy the same lowest energy state. The ground state of calcium would be $1s^{20}$.

EVALUATE Without the Pauli exclusion principle, all ground-state atoms would have all their electrons in the $1s$ state. The world, and the universe, would be a rather boring place, since all atoms would have similar chemical properties.

2: Identify the atom

Determine the element corresponding to the following ground-state electron configurations:

(a) $1s^2 2s^2 2p^6 3s^2 3p^3$

(b) $1s^2 2s^2 2p^6 3s^2 3p^6 4s$

(c) $1s^2 2s^2 2p^6 3s^2 3p^6 3d^3 4s^2$

IDENTIFY, SET UP, AND EXECUTE We could determine the elements in all of the preceding cases by comparing the configurations against Table 3 in the text. A simpler solution is found by counting the electrons in each configuration and finding the element with the correct Z.

Element (a) has 15 electrons and is phosphorus. Element (b) has 19 electrons and is potassium. Element (c) has 23 electrons and is vanadium.

EVALUATE As we become familiar with electron configurations, the problems become easier. Our next step would be to examine excited states of elements. How could we identify the element in those situations? We would follow the same procedure: Count the electrons and correlate with the Z. The excited states contain the same number of electrons, with some of the electrons occupying higher substates.

Problems

1: Possible states of hydrogen

An electron is in the hydrogen atom with $n = 6$. Find the possible values of L and L_z for this electron.

IDENTIFY AND SET UP For $n = 6$, the largest possible value of l is 5 and the largest possible value of m_ℓ is 5. The possible values of L and L_z are found from the angular momentum relations.

EXECUTE The values of L and L_z are given by

$$L = \sqrt{l(l+1)}\, \hbar, \qquad L_z = m_l \hbar.$$

In this problem, l ranges from 0 to 5 and m_l ranges from 0 to ± 5. The values are then

$$
\begin{array}{lll}
L = \sqrt{l(l+1)}\, \hbar = 0, & L_z = m_l \hbar = 0 & (\ell = 0), \\
L = \sqrt{l(l+1)}\, \hbar = \sqrt{1(1+1)}\, \hbar = \sqrt{2}\, \hbar, & L_z = 0, \pm\hbar & (\ell = 1), \\
L = \sqrt{l(l+1)}\, \hbar = \sqrt{2(2+1)}\, \hbar = \sqrt{6}\, \hbar, & L_z = 0, \pm\hbar, \pm 2\hbar & (\ell = 2), \\
L = \sqrt{l(l+1)}\, \hbar = \sqrt{3(3+1)}\, \hbar = \sqrt{12}\, \hbar, & L_z = 0, \pm\hbar, \pm 2\hbar, \pm 3\hbar & (\ell = 3), \\
L = \sqrt{l(l+1)}\, \hbar = \sqrt{4(4+1)}\, \hbar = \sqrt{20}\, \hbar, & L_z = 0, \pm\hbar, \pm 2\hbar, \pm 3\hbar, \pm 4\hbar & (\ell = 4), \\
L = \sqrt{l(l+1)}\, \hbar = \sqrt{5(5+1)}\, \hbar = \sqrt{30}\, \hbar, & L_z = 0, \pm\hbar, \pm 2\hbar, \pm 3\hbar, \pm 4\hbar, \pm 5\hbar & (\ell = 5).
\end{array}
$$

KEY CONCEPT 3 **EVALUATE** As the principal quantum number increases, the number of possible states increases rapidly.

Practice Problem: What is the lowest value of the magnetic quantum number when $n = 4$?
Answer: −3

Extra Practice: How many possible values of L exist for $n = 7$? *Answer:* 7

2: Electron configuration of gallium

Write the ground-state electron configuration for gallium. What next-smaller and next-larger Z's have chemical properties similar to those of gallium?

IDENTIFY AND SET UP Gallium has an atomic number of 31, so we must fill the lowest 31 electron states. Each s subshell can accommodate 2 electrons, each p substate can accommodate 6 electrons, and the d subshells can accommodate 10 electrons.

EXECUTE Gallium's $1s$, $2s$, and $2p$ subshells hold $2 + 2 + 6 = 10$ electrons. The $n = 3$ subshells—$3s$, $3p$, and $3d$—hold $2 + 6 + 10 = 18$ electrons, for a total of 28 electrons in the first 6 subshells. This leaves three electrons. These electrons go into the $4s$ and $4p$ subshells: two in the $4s$ subshell and one in the $4p$ subshell.
   The next-lower Z with chemical properties similar to those of gallium is aluminum since its outer shells are filled with two electrons in the $3s$ subshell and one in the $3p$ subshell. The next-larger Z with chemical properties similar to those of gallium is indium, since its outer shells are filled with two electrons in the $5s$ subshell and one in the $5p$ subshell.

EVALUATE We can see that gallium, aluminum, and indium are chemically similar, since they occupy different rows of the same column in the periodic table.

Practice Problem: What is the next-smaller Z that has similar chemical properties to bromine?
Answer: chlorine

Extra Practice: What is the next-larger Z that has similar chemical properties to germanium?
Answer: tin

3: Zeeman effect

The difference in energies of a hypothetical atom between its $2p$ and $3s$ levels is 1.2 eV. How large a magnetic field would be required to raise the energy of the highest possible state of the $2p$ level to that of the lowest possible $3s$ state due to the electron spin energies?

IDENTIFY AND SET UP The change in energy for a level due to electron spin energies is given by the Zeeman effect. The highest-energy $2p$ state is the energy of the $2p$ state plus the energy difference due to the spin energy. The lowest-energy $3s$ state is the energy of the $3s$ state minus the energy difference due to the spin energy. We'll set these equal to each other to solve the problem.

EXECUTE The energy difference due to a magnetic moment in a magnetic field is given by

$$U = -\vec{\mu} \cdot \vec{B}.$$

The highest-possible-energy $2p$ state in a magnetic field is

$$E_{2p} = E_{2p}(0) + \mu_B B,$$

From Chapter 41 of Student's Study Guide to accompany *University Physics with Modern Physics, Volume 3*, Fifteenth Edition.
Hugh D. Young and Roger A. Freedman.

where $E_{2p}(0)$ is the energy of the 2p state without the electron spin interaction. The lowest-possible-energy 3s state in a magnetic field is

$$E_{3s} = E_{3s}(0) - \mu_B B.$$

We set these energies equal to each other, giving

$$E_{2p} = E_{3s},$$
$$E_{2p}(0) + \mu_B B = E_{3s}(0) - \mu_B B.$$

The energy difference between the 2p and 3s states is 1.2 eV, so we solve for that energy difference:

$$E_{3s}(0) - E_{2p}(0) = 2\mu_B B = 1.2 \text{ eV}.$$

Solving for the magnetic field, we have

$$B = \frac{1.2 \text{ eV}}{2\mu_B} = \frac{(1.2 \text{ eV})(1.6 \times 10^{-19} \text{ J/eV})}{2(9.27 \times 10^{-24} \text{ J/T})} = 1.04 \times 10^4 \text{ T}.$$

KEY CONCEPT 5 **EVALUATE** The magnetic field required is enormous, larger than the largest steady-state field produced by a superconducting magnet. This problem illustrates how electron spin energy is relatively small, although it can be measured (albeit indirectly).

Practice Problem: What is the energy difference between the lowest 2p level and the highest 3p level in this case? *Answer:* 2.4 eV

Extra Practice: What does the energy difference need to be for a strong refrigerator magnet ($B = 100$ G) to raise the energy of the highest 2p state to the lowest 3s state? *Answer:* 1.16 μeV

4: Calculating energy differences from X-ray transitions

The X-ray transitions K_β and L_α are shown in Figure 1. The energies of the X-ray photons emitted in those transitions are shown for five elements in Table 1. Calculate the energy differences between the $n=2$ and $n=1$ levels and the $n=3$ and $n=2$ levels, using the data provided.

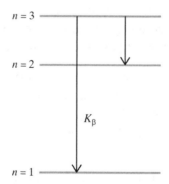

Figure 1 Problem 4.

Element	Z	K_β (keV)	L_α (keV)
Mn	25	6.51	0.721
Zn	30	9.57	1.11
Br	35	13.3	1.60
Zr	40	17.7	2.06
Rh	45	22.8	2.89

Table 1 Problem 4.

IDENTIFY AND SET UP To solve the problem, we'll use the definition of X-ray transitions. The K_β line arises from the energy difference between the $n=3$ and $n=1$ states, so it is equal to $E_3 - E_1$. The L_α line arises from the energy difference between the $n=3$ and $n=2$ states, so it is equal to $E_3 - E_2$. We'll combine these results to find the solution.

EXECUTE The energy difference between the $n=2$ and $n=1$ states is found by combining information from both lines. Specifically,

$$E_2 - E_1 = (E_3 - E_1) - (E_3 - E_2) = E_{K_\beta} - E_{L_\alpha}.$$

The energy difference between the $n=3$ and $n=2$ states is found directly from the L_α line. The results for the energy differences are therefore as follows:

Element	$E_2 - E_1$ (keV)	$E_3 - E_2$ (keV)
Mn	5.79	0.721
Zn	8.46	1.11
Br	11.7	1.60
Zr	15.6	2.06
Rh	19.9	2.89

Table 1 Problem 4.1

KEY CONCEPT 11 EVALUATE Examining the trends in these elements, we see that the energy differences increase with increasing atomic number. What does that tell us about atomic structure?

Try It Yourself!

1: Filling shells

Write down the expected electron configuration for hydrogen atom states (a) for an atom with 18 electrons and (b) for an atom with 22 electrons. (c) If an atom has 22 electrons, how many additional electrons would be required to complete a closed shell?

Solution Checkpoints

IDENTIFY AND SET UP Fill the electron shells from lowest to highest. How many electrons fit in each shell?

EXECUTE (a) Two electrons will fill the $1s$ subshell, and then two electrons will fill the $2s$ subshell. Continue filling until you have

$$1s^2 2s^2 2p^6 3s^2 3p^6$$

for the 18-electron atom.

(b) Starting with the 18 electrons of part (a), add 4 more electrons to the next subshell. This will give

$$1s^2 2s^2 2p^6 3s^2 3p^6 4s^2 3d^2$$

for the 22-electron atom.

(c) The $3d$ subshell holds 10 electrons and the 22-electron atom has 2 atoms in the $3d$ subshell, so 8 additional electrons are required to fill the $3d$ subshell completely.

KEY CONCEPT 3 **EVALUATE** Remember that each subshell hold $2(2l+1)$ electrons.

2: Classical versus quantum physics

(a) Calculate the classical angular precession frequency of an electron in a constant magnetic field. (b) If the electron's angular momentum is quantized, find the allowed values of the possible orbits. (c) For an electron spin in a constant magnetic field, find the difference in energy of its two states and the corresponding angular frequency.

Solution Checkpoints

IDENTIFY AND SET UP The classical precession frequency is found by looking at the forces acting on the electron. Quantized angular momentum requires that the angular momentum have only discrete values.

EXECUTE (a) Classically, an electron is acted upon by a magnetic force in a constant magnetic field and follows a circular path. This gives

$$evB = \frac{mv^2}{R}.$$

The angular frequency is then

$$\omega_c = \frac{eB}{m}.$$

(b) For the angular momentum to be quantized, we must have

$$mvR = n\hbar.$$

Using the results from part (a), we find that the allowed values of the radius are

$$R_n = \sqrt{\frac{n\hbar}{eB}},$$

for integer values of n.

(c) The energy difference between the two states is twice the value of the Zeeman effect:

$$\Delta E = 2\mu_B B.$$

This corresponds to a frequency difference of

$$f = \frac{\Delta E}{h} = \frac{2\mu_B B}{h}.$$

KEY CONCEPT 7 **EVALUATE** How do the two frequencies compare? Check that they are the same by converting the classical angular frequency to ordinary frequency and replacing the Bohr magneton with its definition.

Key Example Variation Problems

Solutions to these problems are in Chapter 41 of the Student's Solutions Manual.

Be sure to review EXAMPLE 41.1 (Section 41.2) before attempting these problems.

VP41.1.1 A particle in the three-dimensional box shown in Fig. 41.1 is in the state $n_X = 2, n_Y = 1, n_Z = 3$. Find (a) the planes (other than the walls of the box) on which the probability distribution function is zero and (b) the probability that the particle will be found somewhere in the region $0 \le x \le L/3$.

VP41.1.2 Half of the volume of the three-dimensional box shown in Fig. 41.1 is in the region $L/4 \le x \le 3L/4$. Find the probability that a particle in the box will be found in this region if the state of the particle is (a) $n_X = 1, n_Y = 1, n_Z = 1$; (b) $n_X = 2, n_Y = 1, n_Z = 2$; (c) $n_X = 3, n_Y = 2, n_Z = 3$; (d) $n_X = 4, n_Y = 1, n_Z = 1$.

VP41.1.3 The region $0 \le x \le L/4, 0 \le y \le L/4$ makes up $\frac{1}{16} = 0.0625$ of the volume of the three-dimensional box shown in Fig. 41.1. Find the probability that a particle in the box will be found in this region if the state of the particle is (a) $n_X = 1, n_Y = 1, n_Z = 1$; (b) $n_X = 2, n_Y = 1, n_Z = 2$; (c) $n_X = 3, n_Y = 2, n_Z = 3$; (d) $n_X = 4, n_Y = 1, n_Z = 1$.

VP41.1.4 Consider the cubical region given by $L/4 \le x \le 3L/4, L/4 \le y \le 3L/4, L/4 \le z \le 3L/4$ at the center of the three-dimensional box shown in Fig. 41.1. (a) What fraction of the total volume of the box is inside this cubical region? (b) If a particle in the box is in the state $n_X = 1, n_Y = 1, n_Z = 1$, find the probability that it will be found somewhere in this cubical region at the center of the box.

Be sure to review EXAMPLES 41.2, 41.3, and 41.4 (Section 41.3) before attempting these problems.

VP41.4.1 Consider the $n = 6$ states of the hydrogen atom. (a) How many distinct (l, m_l) states are there? (b) In terms of $\hbar$, what is the maximum magnitude of the orbital angular momentum L? (c) In terms of $\hbar$, what is the maximum value of the z-component of orbital angular momentum?

VP41.4.2   (a) List all the possible combinations of values of l and m_l for the $n = 3$ states of the hydrogen atom. (b) For which of these states is the angle between the orbital angular momentum vector and the negative z-axis a minimum, and what is that angle?

VP41.4.3   A photon is emitted when a hydrogen atom transitions from one energy level to a lower energy level. Find the energy of this photon, in eV, for each transition:
(a) $n = 3, l = 2, m_l = -2$ to $n = 2, l = 1, m_l = -1$;
(b) $n = 4, l = 2, m_l = 1$ to $n = 2, l = 1, m_l = 0$;
(c) $n = 2, l = 1, m_l = 1$ to $n = 1, l = 0, m_l = 0$.

VP41.4.4   The wave function for an electron in a 1s state in a hydrogen atom is
$\psi_{1s}(r) = 1/\sqrt{\pi a^3}\, e^{-r/a}$, where r is the distance from the nucleus. Find the probability that the electron will be found in the region (a) $0 \leq r \leq 2a$; (b) $0 \leq r \leq 3a$;
(c) $r \geq 4a$.

Be sure to review EXAMPLES 41.6, 41.7, and 41.8 (Section 41.5) before attempting these problems.

VP41.8.1   An isolated electron is placed in a magnetic field $\vec{B} = (3.14\text{ T})\hat{k}$. (a) Find the difference in energy between the $S_z = +\frac{1}{2}\hbar$ and $S_z = -\frac{1}{2}\hbar$ states of the electron. (b) Which state has the higher energy?

VP41.8.2   The outermost electron in a potassium atom is in an $l = 0$ state. If you place a potassium atom in a magnetic field of magnitude 2.36 T and illuminate it with monochromatic electromagnetic radiation, what must be the frequency and wavelength of the radiation to cause a transition between the spin-up and spin-down states of the outermost electron?

VP41.8.3   When the outermost electron in a potassium atom makes a transition from a 4p level to a 4s level, the wavelength of the photon it emits can be either 766.490 nm or 769.896 nm depending on the initial spin orientation of the electron. Find (a) the energy difference between the two 4p levels and (b) the effective magnetic field that the electron experiences in the 4p levels.

VP41.8.4   (a) Find the energy in terms of a state of the electron in a hydrogen atom with $n = 3,\ l = 1,\ j = \frac{3}{2}$ and the energy in terms of a state with $n = 3,\ l = 1,\ j = \frac{1}{2}$. (b) Find the difference between these energies, in eV. Which state has the higher energy? (c) Find the difference in wavelengths of the photons emitted in transitions from each of the states in part (a) to the state $n = 2,\ l = 0,\ j = \frac{1}{2}$. For which initial state is the wavelength longer?

STUDENT'S SOLUTIONS MANUAL FOR ATOMIC STRUCTURE

VP41.1.1. **IDENTIFY:** This problem involves a particle in a three-dimensional cubical box.

SET UP: Eq. (41.15) gives the wave function of the particle, and the square of the wave function is the probability density.

EXECUTE: **(a)** We want to find where the probability distribution function is zero.

$$|\psi_{2,1,3}|^2 = |C|^2 \sin^2 \frac{2\pi x}{L} \sin^2 \frac{\pi y}{L} \sin^2 \frac{3\pi z}{L} = 0.$$

$2\pi x/L = \pi : x = L/2$

$\pi y/L = \pi : y = L$ (none inside the box)

$3\pi z/L = m\pi : z = mL/3 = L/3$ and $2L/3$.

(b) We want the probability that the particle will be found within the range $0 \leq x \leq L/3$. Use results from Example 41.1. $C = (2/L)^{3/2}$. The probability is given by

$$P = |C|^2 \int_0^{L/3} \sin^2 \frac{2\pi x}{L} dx \int_0^L \sin^2 \frac{\pi y}{L} dy \int_0^L \sin^2 \frac{3\pi z}{L} dz.$$

The y and z integrals are each equal to $L/2$, and the x integral is $L/6 - (L/8\pi)\sin 4\pi/3 = 0.201L$. The probability is $P = [(2/L)^{3/2}]^2 (0.201L)(L/2)(L/2) = 0.402$.

EVALUATE: The probability of finding the particle between $x = 0$ and $x = L/3$ is about 40%, so the probability of finding it between $x = L/3$ and L is about 60%.

VP41.1.2. **IDENTIFY:** This problem involves a particle in a three-dimensional cubical box.

SET UP: Eq. (41.15) gives the wave function of the particle, and the square of the wave function is the probability density. We want the probability of finding the particle in the region $L/4 \leq x \leq 3L/4$.

EXECUTE: **(a)** The state is (1,1,1). We integrate using results from Example 41.1. $C = (2/L)^{3/2}$.

$$P = |C|^2 \int_{L/4}^{3L/4} \sin^2 \frac{\pi x}{L} dx \int_0^L \sin^2 \frac{\pi y}{L} dy \int_0^L \sin^2 \frac{\pi z}{L} dz.$$

The second two integrals are each equal to $L/2$, and the first one is $L(1/4 + 1/2\pi)$. Therefore, $P = [(2/L)^{3/2}]^2 (L/2)^2 [L(1/4 + 1/2\pi)] = 1/2 + 1/\pi = 0.818$.

(b) The state is (2,1,2). The only difference from part (a) is the x integral, which is

$$\int_{L/4}^{3L/4} \sin^2 \frac{2\pi x}{L} dx = L/4.$$

Therefore $P = (2/L)^3 (L/2)^2 (L/4) = 0.500$.

(c) The state is (3,2,3). Only the x integral is different from parts (a) and (b).

$$\int_{L/4}^{3L/4} \sin^2 \frac{3\pi x}{L} dx = \frac{L}{4} - \frac{L}{6\pi} = L\left(\frac{1}{4} - \frac{1}{6\pi}\right).$$

$P = (2/L)^3 (L/2)^2 L(1/4 - 1/6\pi) = 0.394$.

(d) The state is (4,1,1). Use the same procedure. In this case, the x integral gives $L/4$, so the probability is $P = (2/L)^3 (L/2)^2 (L/4) = 0.500$.

EVALUATE: Note that the probability is 0.500 whenever n_X is an even integer. Also note that even though the region contains half the volume of the cube, the probability of finding a particle there is not necessarily 0.500.

VP41.1.3. **IDENTIFY:** This problem involves a particle in a three-dimensional cubical box.
SET UP: Eq. (41.15) gives the wave function of the particle, and the square of the wave function is the probability density. We want the probability of finding the particle in the region $0 \leq x \leq L/4$, $0 \leq y \leq L/4$. Use the same approach as in the previous two problems.
EXECUTE: **(a)** The state is (1,1,1). We integrate using results from Example 41.1. $C = (2/L)^{3/2}$.

$$P = \left(\frac{2}{L}\right)^3 \int_0^{L/4} \sin^2 \frac{\pi x}{L} dx \int_0^{L/4} \sin^2 \frac{\pi y}{L} dy \int_0^L \sin^2 \frac{\pi z}{L} dz.$$

The z integral is equal to $L/2$. The y and z integrals are each equal to $L(1/8 - 1/4\pi)$. Therefore
$$P = (2/L)^3 (L)[L(1/8 - 1/4\pi)]^2 = 8.25 \times 10^{-3}.$$
(b) The state is (2,1,2). Use the previous results and those of example 41.1. Doing so, we find that the x integral is $L/8$, the y integral is $L(1/8 - 1/4\pi)$, and the z integral is $L/2$. The probability is
$$P = (2/L)^3 (L/8)[L(1/8 - 1/4\pi)](L/2) = 0.0227.$$
(c) The state is (3,2,3). The x integral is $L(1/8 + 1/12\pi)$, the y integral is $L/8$, the z integral is $L/2$, so
$P = (2/L)^3 [L(1/8 + 1/12\pi)](L/8)(L/2) = 0.0758.$
(d) The state is (4,1,1). x integral is $L/8$, y integral is $L(1/8 - 1/4\pi)$, z integral is $L/2$, so $P = 0.0227$.
EVALUATE: Note that $P_{4,1,1}$ is the same as $P_{2,1,2}$. Also note that even though the region contains 6.25% of the volume of the cube, the probability of finding a particle there is not necessarily 0.0625.

VP41.1.4. **IDENTIFY:** This problem involves a particle in a three-dimensional cubical box.
SET UP: Eq. (41.15) gives the wave function of the particle, and the square of the wave function is the probability density.
EXECUTE: **(a)** We want V_{inside}/V_{box}. $V_{box} = L^3$ and $V_{inside} = (L/2)^3 = L^3/8$. Dividing the volumes gives $V_{inside}/V_{box} = (L^3/8)/L^3 = 1/8 = 0.125$.
(b) We want the probability that the particle is in this region. We use the wave functions in Eq. (41.15), square them and integrate to get the probability as in the previous problems. All the integrals are of the same form as the x integral in problem VP41.1.2(a), so each integral gives the factor $L(1/4 + 1/2\pi)$. The probability is $P = (2/L)^3 [L(1/4 + 1/2\pi)]^3 = 0.548$.
EVALUATE: Note that the probability of finding the particle in the given volume is not necessarily the same as the percent that volume is of the total volume of the box. We have see the same thing in previous problems.

VP41.4.1. **IDENTIFY:** We are looking at the possible states of the hydrogen atom with $n = 3$.
SET UP: $l = 0, 1, 2, ..., n - 1$ and $m_l = 0, \pm 1, \pm 2, ..., \pm l$. $L = \sqrt{l(l+1)} \, \hbar$. $L_z = m_l \hbar$, $\cos \theta_L = L_z / L$.
EXECUTE: **(a)** $l = 0, 1, ..., 5$ and $m_l = 0, \pm 1, \pm 2, ..., \pm 5$. The possible states are:
$l = 0$, $m_l = 0$: 1 state
$l = 1$, $m_l = 0, \pm 1$: 3 states
$l = 2$, $m_l = 0, \pm 1, \pm 2$: 5 states
$l = 3$, $m_l = 0, \pm 1, \pm 2, \pm 3$: 7 states
$l = 4$, $m_l = 0, \pm 1, \pm 2, \pm 3, \pm 4$: 9 states
$l = 5$, $m_l = 0, \pm 1, \pm 2, \pm 3, \pm 4, \pm 5$: 11 states
The total is 36 states.
(b) $l_{max} = 5$, so $L_{max} = \sqrt{5(5+1)} \, \hbar = \sqrt{30} \, \hbar$.
(c) $L_z = m_l \hbar$, so the maximum L_z is $5\hbar$.
EVALUATE: As n increases, the number of possible states increases rapidly.

VP41.4.2. **IDENTIFY:** We are looking at the possible states of the hydrogen atom with $n = 6$.
SET UP: $l = 0, 1, 2, ..., n - 1$ and $m_l = 0, \pm 1, \pm 2, ..., \pm l$. $L = \sqrt{l(l+1)} \, \hbar$. $L_z = m_l \hbar$.

EXECUTE: **(a)** Using the restrictions on l and m_l gives

$l = 0$, $m_l = 0$

$l = 1$, $m_l = 0, \pm 1$

$l = 2$, $m_l = 0, \pm 1, \pm 2$

So the (l, m_l) states are $(0, 0)$, $(1, -1)$, $(1, 0)$, $(1, 1)$, $(2, -2)$, $(2, -1)$, $(2, 0)$, $(2, 1)$, $(2, 2)$.

(b) Using $\cos\theta_L = L_z/L$, we see that the angle between the orbital angular momentum and the *negative z*-axis will be a minimum when θ_L is closest to π, so the cosine should be closest to -1.

$$\cos\theta_L = L_z/L = \frac{m_l\hbar}{\sqrt{l(l+1)}\hbar} = \frac{m_l}{\sqrt{l(l+1)}}.$$

So for $m_l = -2$ and $l = 2$, we get $\cos\theta_L = -2/\sqrt{6}$, which gives $\theta_L = 144.7°$. Therefore the angle with the $-z$-axis is $180° - 144.7° = 35.3°$. This is for the state $l = 2$, $m_l = -2$.

EVALUATE: The greatest angle that the angular momentum makes with the $+z$-axis is $144.7°$.

VP41.4.3. **IDENTIFY:** We are looking at transition energy in the hydrogen atom.

SET UP: $E_n = -(13.60\text{ eV})/n^2$. The energy of a shell depends on n. The photon energy is equal to the energy lost by an electron due to its transition to a lower energy level.

EXECUTE: **(a)** $n = 3 \rightarrow n = 2$. $\Delta E = (13.6\text{ eV})(1/2^2 - 1/3^2) = 1.89\text{ eV}$.

(b) $n = 4 \rightarrow n = 2$. $\Delta E = (13.6\text{ eV})(1/2^2 - 1/4^2) = 2.55\text{ eV}$.

(c) $n = 2 \rightarrow n = 1$. $\Delta E = (13.6\text{ eV})(1/1^2 - 1/2^2) = 10.2\text{ eV}$.

EVALUATE: Note that the energy difference between adjacent shells is not the same but depends on the value of n.

VP41.4.4. **IDENTIFY:** We are investigating the properties of the hydrogen wave function.

SET UP: The probability density is the square of the wave function. We want the probability that the electron will be found within each given region using the given wave function.

$$P = \int |\psi|^2\, dV = \int \frac{1}{\pi a^3}e^{-2r/a}\,4\pi r^2\,dr = -\frac{4}{a^3}\left(\frac{ar^2}{2} + \frac{a^2 r}{2} + \frac{a^3}{4}\right)e^{-2r/a}.$$

EXECUTE: **(a)** $0 \le r \le 2a$. Evaluating the expression above for the limits of the integral gives

$$P = -\frac{4}{a^3}\left[\left(2a^3 + a^3 + \frac{a^3}{4}\right)e^{-4} - \frac{a^3}{4}\right] = -\left(\frac{13}{e^4} - 1\right) = 0.762.$$

(b) $a \le r \le 3a$. Evaluating as in part (a) for the new limits gives $P = 0.615$.

(c) $r \ge 4a$. In this case, the upper limit is infinity. Evaluating as before gives $P = 0.0138$.

EVALUATE: Our results are consistent with Figure 41.8 in the textbook for a $1s$ electron in hydrogen. From the graph for a $1s$ electron, we see that the probability distribution for $P(r)$ is a maximum at $r = a$. The range in part (a) includes that value, so the probability for that range is greater than for the others that do not include it.

VP41.8.1. **IDENTIFY:** This problem involves the Zeeman effect in which an external magnetic field affects the energy levels.

SET UP: The following conditions hold:

$$U = -\mu_z B, \mu_z = -(2.00232)\frac{e}{2m}S_z, \ S_z = \pm\frac{\hbar}{2}.$$

EXECUTE: **(a)** We want the energy differenc. Combing these conditions gives

$$\Delta U = (2.00232)\frac{eB}{2m}\Delta S_z = (2.00232)\frac{eB}{2m}\left[\frac{\hbar}{2} - \left(-\frac{\hbar}{2}\right)\right] = (2.00232)\mu_B B.$$

Using $B = 3.14$ T and $\mu_B = 5.788 \times 10^{-5}$ eV/T gives $\Delta U = 3.64 \times 10^{-4}$ eV.

(b) We want to know which state has greater energy.

$$U = -\mu_z B = -(-2.00232)\frac{e}{2m}S_z B = +(2.00232)\frac{eB}{2m}S_z, \text{ so } S_z = +\frac{\hbar}{2} \text{ has greater energy.}$$

EVALUATE: The *magnitude* of the energy is the same for either of the S_z states.

VP41.8.2. IDENTIFY: We are looking at the effect of an external magnetic field on an atom.
SET UP: The energy of the radiation must be equal to the energy difference between the two spin states of the outermost electron. The following conditions hold:

$$U = -\mu_z B, \mu_z = -(2.00232)\frac{e}{2m}S_z, S_z = \pm\frac{\hbar}{2}$$

EXECUTE: As we saw in problem VP41.8.1(a).

$$\Delta U = (2.00232)\frac{eB}{2m}\Delta S_z = (2.00232)\frac{eB}{2m}\left[\frac{\hbar}{2} - \left(-\frac{\hbar}{2}\right)\right] = (2.00232)\mu_B B.$$

The radiation energy hf must therefore be $hf = (2.00232)\mu_B B$. Using $B = 2.36$ T and solving for f gives $f = 6.61 \times 10^{10}$ Hz. The wavelength is $\lambda = c/f = 4.53$ mm.
EVALUATE: The energy difference between the two spin states is very small, so the wavelength is much longer than that of visible light.

VP41.8.3. IDENTIFY: This problem deals with the effect of a magnetic field on the energy levels of an atom.
SET UP: We know the energy of a photon and from the two previous problems we know the energy difference due to the magnetic field.

$$\Delta U = (2.00232)\mu_B B, E = hc/\lambda$$

EXECUTE: **(a)** We want the energy difference between the two $4p$ levels.

$$\Delta E = \frac{hc}{\lambda_1} - \frac{hc}{\lambda_2} = \frac{hc(\lambda_2 - \lambda_1)}{\lambda_1 \lambda_2}.$$

Using the given wavelengths gives $\Delta E = 7.16 \times 10^{-3}$ eV.
(b) We want the magnetic field. We use the energy difference from part (a) and solve for B.
$$\Delta U = (2.00232)\mu_B B$$

$$B = 61.8 \text{ T.}$$

EVALUATE: This is a *very* strong magnetic field.

VP41.8.4. IDENTIFY: This problem involves the effect of spin and angular momentum on the energy of the hydrogen atom.
SET UP: Eq. (41.41) gives the energy levels, where $j = |l \pm \frac{1}{2}|$.
EXECUTE: **(a)** Use Eq. (41.41) to calculate the energy levels.

$$E_{3,3/2} = -\frac{13.60 \text{ eV}}{3^2}\left[1 + \frac{\alpha^2}{3^2}\left(\frac{3}{3/2 + 1/2} - \frac{3}{4}\right)\right] = -1.511 \text{ eV}\left(1 + \frac{\alpha^2}{12}\right).$$

$$E_{3,1/2} = -1.511 \text{ eV}\left(1 + \frac{\alpha^2}{4}\right)$$

(b) $E_{3,3/2} - E_{3,1/2} = -1.511 \text{ eV}\left[\left(1 + \frac{\alpha^2}{12}\right) - \left(1 + \frac{\alpha^2}{4}\right)\right] = (-1.511\alpha^2)\left(\frac{1}{12} - \frac{1}{4}\right) = 1.34 \times 10^{-5} \text{ eV.}$

The difference is positive, so the $j = 3/2$ state has higher energy.
(c) We want the difference in wavelengths. Use the answer from part (b) and the approach of Example 41.8.
As in that example, the difference in the wavelengths is

$$\Delta\lambda = -\frac{\lambda}{E_{photon}}\Delta E_{photon}.$$

The *difference* in the energy between the $j = 3/2$ and $j = 1/2$ levels of the $n = 3$ shell is *very small* compared to the energy of that shell. Therefore the energy of the photon is essentially $E_3 - E_2$.

$$E_{photon} = E_3 - E_2 = -\frac{13.60 \text{ eV}}{3^2} - \left(-\frac{13.60 \text{ eV}}{2^2}\right) = 1.8890 \text{ eV}.$$

The wavelength of this photon is

$$\lambda = hc/E_{photon} = hc/(1.8890 \text{ eV}) = 6.565 \times 10^{-7} \text{ m}.$$

The difference in the wavelengths is therefore

$$\Delta\lambda = -\frac{\lambda}{E_{photon}}\Delta E_{photon} = -\frac{4.136 \times 10^{-15} \text{ eV} \cdot \text{s}}{1.8890 \text{ eV}}(1.34 \times 10^{-5} \text{ eV}) = -4.66 \times 10^{-12} \text{ m} = -4.66 \times 10^{-3} \text{ nm}.$$

$\lambda_1 - \lambda_2$ is negative, so $\lambda_2 > \lambda_1$ with λ_2 being for the $j = \frac{1}{2}$ initial state.

EVALUATE: As a check, you could also calculate the energy differences using Eq. (41.41) and calculate the wavelength differences from those results. But the algebra gets a bit tedious.

41.5. **IDENTIFY:** This problem is about an electron in a three-dimensional cubical box.

SET UP: The energy levels are given by Eq. (41.16). In the ground state, all three quantum numbers are equal to 1. In the second excited state, two are equal to 2 and one is equal to 1. Therefore the energy levels are $E_1 = \frac{3\pi^2\hbar^2}{2mL^2}$ and $E_2 = \frac{9\pi^2\hbar^2}{2mL^2}$. The photon energy is $E = hc/\lambda$. We want L.

EXECUTE: The energy of the photon is equal to the energy difference between the two states.

$$\Delta E = (9-3)\frac{\pi^2\hbar^2}{2mL^2} = \frac{hc}{\lambda}. \quad L = \sqrt{\frac{3h\lambda}{4mc}} = \sqrt{\frac{3h(8.00 \text{ nm})}{4mc}} = 0.121 \text{ nm}.$$

EVALUATE: The diameter of a Bohr-model hydrogen atom is about 0.11 nm, so this box is around that size.

41.7. **IDENTIFY:** This problem is about an electron in a three-dimensional cubical box.

SET UP: The probability density is the square of the wave function, and Eq. (41.15) gives the wave function. We want the planes for which the probability density is zero, which is where the wave function is zero.

EXECUTE: *x-axis:* $n_X = 3$, so $3\pi x/L = \pi, 2\pi$. This gives $x = L/3$, $x = 2L/3$.
y-axis: $n_Y = 2$, so $2\pi y/L = \pi$. This gives $y = L/2$.
z-axis: $n_Z = 1$, so $\pi z/L = \pi$. This gives $z = L$, so there are no z-planes within the box.

EVALUATE: For the z-axis the probability is zero only at the walls of the box.

41.9. **IDENTIFY:** The possible values of the angular momentum are limited by the value of n.

SET UP: For the N shell $n = 4$, $0 \le l \le n-1$, $|m| \le l$, $m_s = \pm\frac{1}{2}$.

EXECUTE: **(a)** The smallest l is $l = 0$. $L = \sqrt{l(l+1)}\hbar$, so $L_{min} = 0$.

(b) The largest l is $n-1 = 3$, so $L_{max} = \sqrt{3(4)}\hbar = 2\sqrt{3}\hbar = 3.65 \times 10^{-34} \text{ kg} \cdot \text{m}^2/\text{s}$.

(c) Let the chosen direction be the z-axis. The largest m is $m = l = 3$.

$$L_{z,max} = m\hbar = 3\hbar = 3.16 \times 10^{-34} \text{ kg} \cdot \text{m}^2/\text{s}.$$

(d) $S_z = \pm\frac{1}{2}\hbar$. The maximum value is $S_z = \hbar/2 = 5.27 \times 10^{-35} \text{ kg} \cdot \text{m}^2/\text{s}$.

(e) $\frac{S_z}{L_z} = \frac{\frac{1}{2}\hbar}{3\hbar} = \frac{1}{6}$.

EVALUATE: The orbital and spin angular momenta are of comparable sizes.

41.13. **IDENTIFY** and **SET UP:** The smallest nonzero angle for a given l occurs for $m_l = +l$. $L = \sqrt{l(l+1)}\hbar$ and $L_z = m_l\hbar$ where $m_l = 0, \pm 1, \pm 2, ..., \pm l$. $\cos\theta = L_z/L$.

EXECUTE: In this case $\theta = 26.6°$, so $\cos 26.6° = \dfrac{l}{\sqrt{l(l+1)}}$. Squaring gives $l(l+1)\cos^2(26.6°) = l^2$.

Solving for l gives $l = \dfrac{\cos^2(26.6°)}{1 - \cos^2(26.6°)} = 4$.

EVALUATE: For $l = 4$ we see that the angle between the angular momentum vector and the positive z-axis ranges from $26.6°$ ($m_l = +l$) to $180° - 26.6° = 153.4°$ ($m_l = -l$).

41.17. **IDENTIFY:** For the 5g state, $l = 4$, which limits the other quantum numbers.

SET UP: $m_l = 0, \pm 1, \pm 2, ..., \pm l$. g means $l = 4$. $\cos\theta = L_z/L$, with $L = \sqrt{l(l+1)}\,\hbar$ and $L_z = m_l\hbar$.

EXECUTE: **(a)** There are eighteen 5g states: $m_l = 0, \pm 1, \pm 2, \pm 3, \pm 4$, with $m_s = \pm\frac{1}{2}$ for each.

(b) The largest θ is for the most negative m_l. $L = 2\sqrt{5}\hbar$. The most negative L_z is $L_z = -4\hbar$.

$$\cos\theta = \frac{-4\hbar}{2\sqrt{5}\hbar} \text{ and } \theta = 153.4°.$$

(c) The smallest θ is for the largest positive m_l, which is $m_l = +4$. $\cos\theta = \dfrac{4\hbar}{2\sqrt{5}\hbar}$ and $\theta = 26.6°$.

EVALUATE: The minimum angle between $\vec{L}$ and the z-axis is for $m_l = +l$ and for that m_l,

$$\cos\theta = \frac{l}{\sqrt{l(l+1)}}.$$

41.21. **IDENTIFY:** Apply $\Delta U = \mu_B B$.

SET UP: For a 3p state, $l = 1$ and $m_l = 0, \pm 1$.

EXECUTE: **(a)** $B = \dfrac{U}{\mu_B} = \dfrac{(2.71 \times 10^{-5} \text{ eV})}{(5.79 \times 10^{-5} \text{ eV/T})} = 0.468$ T.

(b) Three: $m_l = 0, \pm 1$.

EVALUATE: The $m_l = +1$ level will be highest in energy and the $m_l = -1$ level will be lowest. The $m_l = 0$ level is unaffected by the magnetic field.

41.23. **IDENTIFY** and **SET UP:** The interaction energy between an external magnetic field and the orbital angular momentum of the atom is given by $U = m_l \mu_B B$. The energy depends on m_l with the most negative m_l value having the lowest energy.

EXECUTE: **(a)** For the 5g level, $l = 4$ and there are $2l + 1 = 9$ different m_l states. The 5g level is split into 9 levels by the magnetic field.

(b) Each m_l level is shifted in energy an amount given by $U = m_l \mu_B B$. Adjacent levels differ in m_l by one, so $\Delta U = \mu_B B$.

$$\mu_B = \frac{e\hbar}{2m} = \frac{(1.602 \times 10^{-19} \text{ C})(1.055 \times 10^{-34} \text{ J} \cdot \text{s})}{2(9.109 \times 10^{-31} \text{ kg})} = 9.277 \times 10^{-24} \text{ A} \cdot \text{m}^2.$$

$$\Delta U = \mu_B B = (9.277 \times 10^{-24} \text{ A/m}^2)(0.600 \text{ T}) = 5.566 \times 10^{-24} \text{ J}(1 \text{ eV}/1.602 \times 10^{-19} \text{ J}) = 3.47 \times 10^{-5} \text{ eV}.$$

(c) The level of highest energy is for the largest m_l, which is $m_l = l = 4$; $U_4 = 4\mu_B B$. The level of lowest energy is for the smallest m_l, which is $m_l = -l = -4$; $U_{-4} = -4\mu_B B$. The separation between these two levels is $U_4 - U_{-4} = 8\mu_B B = 8(3.47 \times 10^{-5} \text{ eV}) = 2.78 \times 10^{-4} \text{ eV}$.

EVALUATE: The energy separations are proportional to the magnetic field. The energy of the $n = 5$ level in the absence of the external magnetic field is $(-13.6 \text{ eV})/5^2 = -0.544 \text{ eV}$, so the interaction energy with the magnetic field is much less than the binding energy of the state.

41.27. **IDENTIFY and SET UP:** The interaction energy is $U = -\vec{\mu} \cdot \vec{B}$, with μ_z given by

$$\mu_z = -(2.00232)\left(\frac{e}{2m}\right)S_z.$$

EXECUTE: $U = -\vec{\mu} \cdot \vec{B} = +\mu_z B$, since the magnetic field is in the negative z-direction.

$$\mu_z = -(2.00232)\left(\frac{e}{2m}\right)S_z, \text{ so } U = -(2.00232)\left(\frac{e}{2m}\right)S_z B.$$

$$S_z = m_s \hbar, \text{ so } U = -2.00232\left(\frac{e\hbar}{2m}\right)m_s B.$$

$$\frac{e\hbar}{2m} = \mu_B = 5.788 \times 10^{-5} \text{ eV/T}.$$

$$U = -2.00232\mu_B m_s B.$$

The $m_s = +\frac{1}{2}$ level has lower energy.

$$\Delta U = U(m_s = -\tfrac{1}{2}) - U(m_s = +\tfrac{1}{2}) = -2.00232\,\mu_B B(-\tfrac{1}{2} - (+\tfrac{1}{2})) = +2.00232\,\mu_B B.$$

$$\Delta U = +2.00232(5.788 \times 10^{-5} \text{ eV/T})(1.45 \text{ T}) = 1.68 \times 10^{-4} \text{ eV}.$$

EVALUATE: The interaction energy with the electron spin is the same order of magnitude as the interaction energy with the orbital angular momentum for states with $m_l \neq 0$. But a $1s$ state has $l = 0$ and $m_l = 0$, so there is no orbital magnetic interaction.

41.31. **IDENTIFY:** Write out the electron configuration for ground-state beryllium.

SET UP: Beryllium has 4 electrons.

EXECUTE: **(a)** $1s^2 2s^2$.

(b) $1s^2 2s^2 2p^6 3s^2$. $Z = 12$ and the element is magnesium.

(c) $1s^2 2s^2 2p^6 3s^2 3p^6 4s^2$. $Z = 20$ and the element is calcium.

EVALUATE: Beryllium, calcium, and magnesium are all in the same column of the periodic table.

41.33. **IDENTIFY and SET UP:** The energy of an atomic level is given in terms of n and Z_{eff} by

$$E_n = -\left(\frac{Z_{\text{eff}}^2}{n^2}\right)(13.6 \text{ eV}). \text{ The ionization energy for a level with energy } -E_n \text{ is } +E_n.$$

EXECUTE: $n = 5$ and $Z_{\text{eff}} = 2.771$ gives $E_5 = -\dfrac{(2.771)^2}{5^2}(13.6 \text{ eV}) = -4.18 \text{ eV}$.

The ionization energy is 4.18 eV.

EVALUATE: The energy of an atomic state is proportional to Z_{eff}^2.

41.35. **IDENTIFY** and **SET UP:** Use the exclusion principle to determine the ground-state electron configuration, as in Table 41.3 in the textbook. Estimate the energy by estimating Z_{eff}, taking into account the electron screening of the nucleus.

EXECUTE: **(a)** $Z = 7$ for nitrogen so a nitrogen atom has 7 electrons. N^{2+} has 5 electrons: $1s^2 2s^2 2p$.

(b) $Z_{eff} = 7 - 4 = 3$ for the $2p$ level.

$$E_n = -\left(\frac{Z_{eff}^2}{n^2}\right)(13.6 \text{ eV}) = -\frac{3^2}{2^2}(13.6 \text{ eV}) = -30.6 \text{ eV}.$$

(c) $Z = 15$ for phosphorus so a phosphorus atom has 15 electrons.
P^{2+} has 13 electrons: $1s^2 2s^2 2p^6 3s^2 3p$.

(d) $Z_{eff} = 15 - 12 = 3$ for the $3p$ level.

$$E_n = -\left(\frac{Z_{eff}^2}{n^2}\right)(13.6 \text{ eV}) = -\frac{3^2}{3^2}(13.6 \text{ eV}) = -13.6 \text{ eV}.$$

EVALUATE: In these ions there is one electron outside filled subshells, so it is a reasonable approximation to assume full screening by these inner-subshell electrons.

41.37. **IDENTIFY** and **SET UP:** Apply $E_{K\alpha} \cong (Z-1)^2 (10.2 \text{ eV})$. $E = hf$ and $c = f\lambda$.

EXECUTE: **(a)** $Z = 20$: $f = (2.48 \times 10^{15} \text{ Hz})(20-1)^2 = 8.95 \times 10^{17} \text{ Hz}$.

$E = hf = (4.14 \times 10^{-15} \text{ eV} \cdot \text{s})(8.95 \times 10^{17} \text{ Hz}) = 3.71 \text{ keV}$. $\lambda = \dfrac{c}{f} = \dfrac{3.00 \times 10^8 \text{ m/s}}{8.95 \times 10^{17} \text{ Hz}} = 3.35 \times 10^{-10}$ m.

(b) $Z = 27$: $f = 1.68 \times 10^{18}$ Hz. $E = 6.96$ keV. $\lambda = 1.79 \times 10^{-10}$ m.

(c) $Z = 48$: $f = 5.48 \times 10^{18}$ Hz, $E = 22.7$ keV, $\lambda = 5.47 \times 10^{-11}$ m.

EVALUATE: f and E increase and λ decreases as Z increases.

41.39. **IDENTIFY:** The electrons cannot all be in the same state in a cubical box.

SET UP and **EXECUTE:** The ground state can hold 2 electrons, the first excited state can hold 6 electrons, and the second excited state can hold 6. Therefore, two electrons will be in the second excited state, which has energy $3E_{1,1,1}$.

EVALUATE: The second excited state is the third state, which has energy $3E_{1,1,1}$, as shown in Figure 41.4 in the textbook.

41.41. **IDENTIFY:** Calculate the probability of finding a particle in a given region within a cubical box.

(a) SET UP and **EXECUTE:** The box has volume L^3. The specified cubical space has volume $(L/4)^3$. Its fraction of the total volume is $\dfrac{1}{64} = 0.0156$.

(b) SET UP and **EXECUTE:** $P = \left(\dfrac{2}{L}\right)^3 \left[\int_0^{L/4} \sin^2 \dfrac{\pi x}{L} \, dx\right]\left[\int_0^{L/4} \sin^2 \dfrac{\pi y}{L} \, dy\right]\left[\int_0^{L/4} \sin^2 \dfrac{\pi z}{L} \, dz\right].$

From Example 41.1, each of the three integrals equals $\dfrac{L}{8} - \dfrac{L}{4\pi} = \dfrac{1}{2}\left(\dfrac{L}{2}\right)\left(\dfrac{1}{2} - \dfrac{1}{\pi}\right).$

$$P = \left(\dfrac{2}{L}\right)^3 \left(\dfrac{L}{2}\right)^3 \left(\dfrac{1}{2}\right)^3 \left(\dfrac{1}{2} - \dfrac{1}{\pi}\right)^3 = 7.50 \times 10^{-4}.$$

EVALUATE: Note that this is the cube of the probability of finding the particle anywhere between $x = 0$ and $x = L/4$. This probability is much less that the fraction of the total volume that this space represents. In this quantum state the probability distribution function is much larger near the center of the box than near its walls.

(c) SET UP and EXECUTE: $|\psi_{2,1,1}|^2 = \left(\dfrac{L}{2}\right)^3 \left(\sin^2 \dfrac{2\pi x}{L}\right)\left(\sin^2 \dfrac{\pi y}{L}\right)\left(\sin^2 \dfrac{\pi z}{L}\right)$.

$$P = \left(\frac{2}{L}\right)^3 \left[\int_0^{L/4} \sin^2 \frac{2\pi x}{L} dx\right]\left[\int_0^{L/4} \sin^2 \frac{\pi y}{L} dy\right]\left[\int_0^{L/4} \sin^2 \frac{\pi z}{L} dz\right].$$

$$\left[\int_0^{L/4} \sin^2 \frac{\pi y}{L} dy\right] = \left[\int_0^{L/4} \sin^2 \frac{\pi z}{L} dz\right] = \frac{L}{2}\left(\frac{1}{2}\right)\left(\frac{1}{2} - \frac{1}{\pi}\right). \quad \int_0^{L/4} \sin^2 \frac{2\pi x}{L} dx = \frac{L}{8}.$$

$$P = \left(\frac{2}{L}\right)^3 \left(\frac{L}{2}\right)^2 \left(\frac{1}{2}\right)^2 \left(\frac{1}{2} - \frac{1}{\pi}\right)^2 \left(\frac{L}{8}\right) = 2.06\times10^{-3}.$$

EVALUATE: This is about a factor of three larger than the probability when the particle is in the ground state.

41.45. **IDENTIFY:** Find solutions to Eq. (41.5).

SET UP: $\omega_1 = \sqrt{k_1'/m}$, $\omega_2 = \sqrt{k_2'/m}$. Let $\psi_{n_x}(x)$ be a solution of Eq. (40.44) with $E_{n_x} = (n_x + \frac{1}{2})\hbar\omega_1$, $\psi_{n_y}(y)$ be a similar solution, and let $\psi_{n_z}(z)$ be a solution of Eq. (40.44) but with z as the independent variable instead of x, and energy $E_{n_z} = (n_z + \frac{1}{2})\hbar\omega_2$.

EXECUTE: **(a)** As in Problem 41.44, look for a solution of the form $\psi(x, y, z) = \psi_{n_x}(x)\psi_{n_y}(y)\psi_{n_z}(z)$. Then, $-\dfrac{\hbar^2}{2m}\dfrac{\partial^2\psi}{\partial x^2} = (E_{n_x} - \frac{1}{2}k_1'x^2)\psi$ with similar relations for $\dfrac{\partial^2\psi}{\partial y^2}$ and $\dfrac{\partial^2\psi}{\partial z^2}$. Adding,

$$-\frac{\hbar^2}{2m}\left(\frac{\partial^2\psi}{\partial x^2} + \frac{\partial^2\psi}{\partial y^2} + \frac{\partial^2\psi}{\partial z^2}\right) = (E_{n_x} + E_{n_y} + E_{n_z} - \tfrac{1}{2}k_1'x^2 - \tfrac{1}{2}k_1'y^2 - \tfrac{1}{2}k_2'z^2)\psi$$

$$= (E_{n_x} + E_{n_y} + E_{n_z} - U)\psi = (E - U)\psi$$

where the energy E is $E = E_{n_x} + E_{n_y} + E_{n_z} = \hbar\left[(n_x + n_y + 1)\omega_1^2 + (n_z + \frac{1}{2})\omega_2^2\right]$, with n_x, n_y, and n_z all nonnegative integers.

(b) The ground level corresponds to $n_x = n_y = n_z = 0$, and $E = \hbar(\omega_1^2 + \frac{1}{2}\omega_2^2)$. The first excited level corresponds to $n_x = n_y = 0$ and $n_z = 1$, since $\omega_1^2 > \omega_2^2$, and $E = \hbar(\omega_1^2 + \frac{3}{2}\omega_2^2)$.

(c) There is only one set of quantum numbers for both the ground state and the first excited state.

EVALUATE: For the isotropic oscillator of Problem 41.44 there are three states for the first excited level but only one for the anisotropic oscillator.

41.49. **(a) IDENTIFY and SET UP:** The energy is given by $E_n = K_n + U_n = -\dfrac{1}{\epsilon_0^2}\dfrac{me^4}{8n^2h^2}$ from Chapter 39, which is identical to the Bohr energy levels for hydrogen from this chapter. The potential energy is given by $U(r) = \dfrac{1}{4\pi\epsilon_0}\dfrac{q_1q_2}{r}$, with $q_1 = +Ze$ and $q_2 = -e$.

EXECUTE: $E_{1s} = -\dfrac{1}{(4\pi\epsilon_0)^2}\dfrac{me^4}{2\hbar^2}$; $U(r) = -\dfrac{1}{4\pi\epsilon_0}\dfrac{e^2}{r}$.

$E_{1s} = U(r)$ gives $-\dfrac{1}{(4\pi\epsilon_0)^2}\dfrac{me^4}{2\hbar^2} = -\dfrac{1}{4\pi\epsilon_0}\dfrac{e^2}{r}$.

$$r = \dfrac{(4\pi\epsilon_0)2\hbar^2}{me^2} = 2a.$$

EVALUATE: The turning point is twice the Bohr radius.

(b) IDENTIFY and SET UP: For the 1s state the probability that the electron is in the classically forbidden region is $P(r > 2a) = \int_{2a}^{\infty}|\psi_{1s}|^2\,dV = 4\pi\int_{2a}^{\infty}|\psi_{1s}|^2\,r^2\,dr$. The normalized wave function of the

1s state of hydrogen is given in Example 41.4: $\psi_{1s}(r) = \dfrac{1}{\sqrt{\pi a^3}}e^{-r/a}$. Evaluate the integral; the integrand is the same as in Example 41.4.

EXECUTE: $P(r > 2a) = 4\pi\left(\dfrac{1}{\pi a^3}\right)\int_{2a}^{\infty}r^2 e^{-2r/a}\,dr$.

Use the integral formula $\int r^2 e^{-\alpha r}\,dr = -e^{-\alpha r}\left(\dfrac{r^2}{\alpha} + \dfrac{2r}{\alpha^2} + \dfrac{2}{\alpha^3}\right)$, with $\alpha = 2/a$.

$$P(r > 2a) = -\dfrac{4}{a^3}\left[e^{-2r/a}\left(\dfrac{ar^2}{2} + \dfrac{a^2 r}{2} + \dfrac{a^3}{4}\right)\right]_{2a}^{\infty} = +\dfrac{4}{a^3}e^{-4}(2a^3 + a^3 + a^3/4).$$

$$P(r > 2a) = 4e^{-4}(13/4) = 13e^{-4} = 0.238.$$

EVALUATE: These is a 23.8% probability of the electron being found in the classically forbidden region, where classically its kinetic energy would be negative.

41.53. IDENTIFY: This problem is about the broadening of spectral lines in hydrogen due to spin-orbit coupling.

SET UP: The blue spectral line of wavelength 434 nm is due to a transition from the $n = 5$ to the $n = 2$ shell. Eq. (41.41) applies, with $l = 0, 1, 2, 3, 4$ and $j = |l \pm \frac{1}{2}|$. In any transition $\Delta l = \pm 1$. Use Example 41.8 as a guide.

EXECUTE: (a) We want the number of states.

For the $n = 2$ level: $l = 0$ and 1. For $l = 0$, $j = |0 \pm \frac{1}{2}| = \frac{1}{2}$. For $l = 1$, $j = 1 \pm \frac{1}{2} = 1/2$ and $3/2$. The possible states are $(0, 1/2)$, $(1, 1/2)$, and $(1, 3/2)$, so there are 3 states.

For the $n = 5$ level: $l = 0, 1, 2, 3,$ or 4. The possible values of j are

$l = 0: j = 1/2$
$l = 1: j = 1/2, 3/2$
$l = 2: j = 3/2, 5/2$
$l = 3: j = 5/2, 7/2$
$l = 4: j = 7/2, 9/2$

The possible states are $(0, 1/2)$, $(1, 1/2)$, $(1, 3/2)$, $(2, 3/2)$, $(2, 5/2)$, $(3, 5/2)$, $(3, 7/2)$, $(4, 7/2)$, $(4, 9/2)$. There are a total of 9 states.

(b) We want the number of different energy levels for each value of n. From Eq. (41.41) we see that for a given n, the energy $E_{n,j}$ depends only on j. Therefore all states having the same n and j have the same energy.

$n = 2$ level: Since j has only 2 different possible values (1/2 and 3/2), there are only 2 different energy levels.

$n = 5$ level: Since j has 5 different values (1/2, 3/2, 5/2, 7/2, 9/2) there are 5 different energy levels.

For each energy level, we want its *difference* from $-(13.6 \text{ eV})/n^2$. Call $E_n = -(13.6 \text{ eV})/n^2$ and call the energy difference we want $\Delta E_{n,j} = E_{n,j} - E_n$, where $E_{n,j}$ is the energy of the (n, j) level as given in Eq. (41.41). For calculations use $\alpha^2 = 5.32514 \times 10^{-5}$. Using Eq. (41.41), we get

$$\Delta E_{n,j} = \frac{13.60 \text{ eV}}{n^2}\left[1 + \frac{\alpha^2}{n^2}\left(\frac{n}{j+1/2} - \frac{3}{4}\right)\right] - \left(\frac{13.60 \text{ eV}}{n^2}\right) = -\frac{13.60 \text{ eV}}{n^2}\frac{\alpha^2}{n^2}\left(\frac{n}{j+1/2} - \frac{3}{4}\right).$$

Simplifying to a working equation gives

$$\Delta E_{n,j} = -\frac{13.60 \text{ eV}\,\alpha^2}{n^4}\left(\frac{n}{j+1/2} - \frac{3}{4}\right).$$

We now use this equation to calculate the energy *differences*. If $\Delta E_{n,j}$ is negative, the energy of the level is *less than* E_n.

$\underline{n = 2 \text{ level}}$: For $n = 2$ and $j = 1/2$, we have

$$\Delta E_{2,1/2} = -\frac{13.60 \text{ eV}\,\alpha^2}{2^4}\left(\frac{2}{1/2+1/2} - \frac{3}{4}\right) = -5.658 \times 10^{-5} \text{ eV}.$$

For $n = 2$ and $j = 3/2$, we have

$$\Delta E_{2,3/2} = -\frac{13.60 \text{ eV}\,\alpha^2}{2^4}\left(\frac{2}{3/2+1/2} - \frac{3}{4}\right) = -1.1316 \times 10^{-5} \text{ eV}.$$

$\underline{n = 5 \text{ level}}$: For $n = 5$ and $j = 1/2$, we have

$$\Delta E_{5,1/2} = -\frac{13.60 \text{ eV}\,\alpha^2}{5^4}\left(\frac{5}{1/2+1/2} - \frac{3}{4}\right) = -4.11356 \times 10^{-5} \text{ eV}.$$

For $n = 5$ and $j = 3/2$, we have

$$\Delta E_{5,3/2} = -\frac{13.60 \text{ eV}\,\alpha^2}{5^4}\left(\frac{5}{3/2+1/2} - \frac{3}{4}\right) = -1.69382 \times 10^{-5} \text{ eV}.$$

For $n = 5$ and $j = 5/2$, we have

$$\Delta E_{5,5/2} = -\frac{13.60 \text{ eV}\,\alpha^2}{5^4}\left(\frac{5}{5/2+1/2} - \frac{3}{4}\right) = -8.87239 \times 10^{-6} \text{ eV}.$$

For $n = 5, j = 7/2$, we have

$$\Delta E_{5,7/2} = -\frac{13.60 \text{ eV}\,\alpha^2}{5^4}\left(\frac{5}{7/2+1/2} - \frac{3}{4}\right) = -4.83949 \times 10^{-6} \text{ eV}.$$

For $n = 5, j = 9/2$, we have

$$\Delta E_{5,9/2} = -\frac{13.60 \text{ eV}\,\alpha^2}{5^4}\left(\frac{5}{9/2+1/2} - \frac{3}{4}\right) = -2.41974 \times 10^{-6} \text{ eV}.$$

Notice that in every case, fine structure due to spin-orbit coupling makes the energy levels *more negative*, i.e. *lower*.

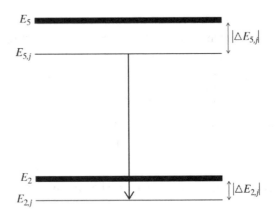

Figure 41.53

(c) We want the transition that emits the *shortest* wavelength of light. The shortest wavelength photon will have the highest energy, so it is between energy levels that have the *greatest* energy *difference* between them. Figure 41.53 illustrates a transition from a $(5, j)$ level to a $(2, j)$ level, where the two values of j are *not* the same. For the energy difference between these two levels to be as large as possible, $|\Delta E_{5,j}|$ should be the *smallest* it can be, and $|\Delta E_{2,j}|$ should be the *largest* it can be. But the transition must also obey $\Delta l = \pm 1$. For the $n = 2$ level, $l = 0$ or 1, which means that in the $n = 5$ level, $l = 0$, 1, or 2. The energy levels, for a given n, depend only on j, so the j values for possible transitions that result in an energy difference and obey $\Delta l = \pm 1$ are

$$j = 1/2 \;\rightarrow\; j = 3/2 \quad (l = 0 \;\rightarrow\; l = 1)$$
$$j = 3/2 \;\rightarrow\; j = 1/2 \quad (l = 1 \;\rightarrow\; l = 0)$$
$$j = 5/2 \;\rightarrow\; j = 1/2 \quad (l = 2 \;\rightarrow\; l = 1)$$
$$j = 5/2 \;\rightarrow\; j = 3/2 \quad (l = 2 \;\rightarrow\; l = 1)$$

Using our results from part (b), we see that $|\Delta E_{2,j}|$ is greatest when $j = 1/2$ ($l = 1$), and $|\Delta E_{5,j}|$ is smallest when $j = 5/2$ ($l = 2$). Therefore the transition from the (5, 2, 5/2) level to the (2, 1, 1/2) level emits a photon of the greatest energy and therefore the shortest wavelength. The energy of this photon is

$E_{photon} = E_5 + \Delta E_{5,5/2} - (E_2 + \Delta E_{2,1/2})$. Using this condition with $E_n = -(13.60 \text{ eV})/n^2$ and our results from part (b), we have

$$E_{photon} = E_5 - E_2 + \Delta E_{5,5/2} - \Delta E_{2,1/2}$$

$$E_{photon} = -\frac{13.60 \text{ eV}}{5^2} - \left(-\frac{13.60 \text{ eV}}{2^2}\right) + (-8.87239 \times 10^{-5} \text{ eV}) - (-5.658 \times 10^{-5} \text{ eV}).$$

$$E_{photon} = \underbrace{2.856 \text{ eV}}_{E_5 \rightarrow E_2} + \underbrace{4.7708 \times 10^{-5} \text{ eV}}_{\Delta E_{5/2 \rightarrow 1/2}}.$$

If there were no spin-orbit coupling, the wavelength of the photon would be 434 nm, as given in the problem. But, due to spin-orbit coupling, what appears as a single spectral line is made up of a number of lines, each of slightly different wavelength. To find the difference of the shortest-wavelength line from 434 nm, we use the approach in Example 41.8. The energy of the photon is $E = hc/\lambda$. As in the example, we can express this as

$$\Delta E = -\frac{hc}{\lambda^2} \Delta\lambda$$

$$\Delta\lambda = \frac{\lambda^2 \Delta E}{hc} = -\frac{(434 \text{ nm})^2 (4.7708 \times 10^{-5} \text{ eV})}{(4.13567 \times 10^{-15} \text{ eV} \cdot \text{s})(2.998 \times 10^8 \text{ m/s})} = -7.26 \times 10^{-12} \text{ m} = -7.26 \times 10^{-3} \text{ nm}.$$

(d) We want the longest wavelength photon. Refer to Figure 41.53. The longest wavelength is due to a transition between levels having the smallest energy difference between them. Using the allowed transitions from part (c), we see that the energy difference is *smallest* when $|\Delta E_{5,j}|$ is a *maximum* and $|E_{2,j}|$ is a *minimum*. For the allowed transitions, we see that $|E_{2,j}|$ is smallest when $j = 3/2$ ($l = 1$), and $|\Delta E_{5,j}|$ is largest when $j = 1/2$ ($l = 0$). So this transition is from the (5, 0, 1/2) level to the (2, 1, 3/2) level. Follow the same approach as for part (c), which gives

$$E_{\text{photon}} = E_5 - E_2 + \Delta E_{5,1/2} - \Delta E_{2,3/2}$$

$$E_{\text{photon}} = -\frac{13.60\,\text{eV}}{5^2} - \left(-\frac{13.60\,\text{eV}}{2^2}\right) + (-4.11356\times10^{-5}\,\text{eV}) - (-1.1316\times10^{-5}\,\text{eV}).$$

$$E_{\text{photon}} = \underbrace{2.856\,\text{eV}}_{E_5 \rightarrow E_2} - \underbrace{2.98196\times10^{-5}\,\text{eV}}_{\Delta E_{1/2 \rightarrow 3/2}}.$$

As before, we have

$$\Delta\lambda = -\frac{\lambda^2 \Delta E}{hc} = -\frac{(434\,\text{nm})^2(-2.98196\times10^{-5}\,\text{eV})}{(4.13567\times10^{-15}\,\text{eV.s})(2.998\times10^8\,\text{m/s})} = +4.535\times10^{-12}\,\text{m} = +4.54\times10^{-3}\,\text{nm}.$$

(e) We want the total broadening of the spectral line. The width is the sum of the quantities we calculated in parts (c) and (d). The width is 7.26×10^{-3} nm + 4.54×10^{-3} nm = 1.18×10^{-2} nm.
EVALUATE: Notice that fine structure effects make vary small changes in the wavelength. That is why they are called *fine* structure.

41.55. **IDENTIFY:** The presence of an external magnetic field shifts the energy levels up or down, depending upon the value of m_l.

SET UP: The energy difference due to the magnetic field is $\Delta E = \mu_B B$ and the energy of a photon is $E = hc/\lambda$.

EXECUTE: For the p state, $m_l = 0$ or ± 1, and for the s state $m_l = 0$. Between any two adjacent lines, $\Delta E = \mu_B B$. Since the change in the wavelength ($\Delta\lambda$) is very small, the energy change (ΔE) is also very small, so we can use differentials. $E = hc/\lambda$. $|dE| = \frac{hc}{\lambda^2}d\lambda$ and $\Delta E = \frac{hc\Delta\lambda}{\lambda^2}$. Since

$\Delta E = \mu_B B$, we get $\mu_B B = \frac{hc\Delta\lambda}{\lambda^2}$ and $B = \frac{hc\Delta\lambda}{\mu_B\lambda^2}$.

$B = (4.136\times10^{-15}\,\text{eV}\cdot\text{s})(3.00\times10^8\,\text{m/s})(0.0462\,\text{nm})/(5.788\times10^{-5}\,\text{eV/T})(575.050\,\text{nm})^2 = 3.00\,\text{T}$.
EVALUATE: Even a strong magnetic field produces small changes in the energy levels, and hence in the wavelengths of the emitted light.

41.59. **IDENTIFY and SET UP:** m_s can take on four different values: $m_s = -\frac{3}{2}, -\frac{1}{2}, +\frac{1}{2}, +\frac{3}{2}$. Each nlm_l state can have four electrons, each with one of the four different m_s values. Apply the exclusion principle to determine the electron configurations.

EXECUTE: **(a)** For a filled $n = 1$ shell, the electron configuration would be $1s^4$; four electrons and $Z = 4$. For a filled $n = 2$ shell, the electron configuration would be $1s^4 2s^4 2p^{12}$; twenty electrons and $Z = 20$.
(b) Sodium has $Z = 11$; eleven electrons. The ground-state electron configuration would be $1s^4 2s^4 2p^3$.
EVALUATE: The chemical properties of each element would be very different.

41.61. **(a) IDENTIFY and SET UP:** The energy of the photon equals the transition energy of the atom: $\Delta E = hc/\lambda$. The energies of the states are given by $E_n = -\dfrac{13.60 \text{ eV}}{n^2}$.

EXECUTE: $E_n = -\dfrac{13.60 \text{ eV}}{n^2}$, so $E_2 = -\dfrac{13.60 \text{ eV}}{4}$ and $E_1 = -\dfrac{13.60 \text{ eV}}{1}$. Thus $\Delta E = E_2 - E_1$ gives

$\Delta E = 13.60 \text{ eV}(-\tfrac{1}{4}+1) = \tfrac{3}{4}(13.60 \text{ eV}) = 10.20 \text{ eV} = (10.20 \text{ eV})(1.602 \times 10^{-19} \text{ J/eV}) = 1.634 \times 10^{-18} \text{ J}.$

$\lambda = \dfrac{hc}{\Delta E} = \dfrac{(6.626 \times 10^{-34} \text{ J} \cdot \text{s})(2.998 \times 10^8 \text{ m/s})}{1.634 \times 10^{-18} \text{ J}} = 1.22 \times 10^{-7} \text{ m} = 122 \text{ nm}.$

(b) IDENTIFY and SET UP: Calculate the change in ΔE due to the orbital magnetic interaction energy, $U = m_l \mu_B B$, and relate this to the shift $\Delta \lambda$ in the photon wavelength.

EXECUTE: The shift of a level due to the energy of interaction with the magnetic field in the z-direction is $U = m_l \mu_B B$. The ground state has $m_l = 0$, so is unaffected by the magnetic field. The $n = 2$ initial state has $m_l = -1$, so its energy is shifted downward an amount

$U = m_l \mu_B B = (-1)(9.274 \times 10^{-24} \text{ A/m}^2)(2.20 \text{ T}) = (-2.040 \times 10^{-23} \text{ J})(1 \text{ eV}/1.602 \times 10^{-19} \text{ J})$

$= 1.273 \times 10^{-4} \text{ eV}.$

Note that the shift in energy due to the magnetic field is a very small fraction of the 10.2 eV transition energy. Problem 39.74c shows that in this situation $|\Delta \lambda / \lambda| = |\Delta E / E|$. This gives

$|\Delta \lambda| = \lambda |\Delta E / E| = 122 \text{ nm} \left(\dfrac{1.273 \times 10^{-4} \text{ eV}}{10.2 \text{ eV}} \right) = 1.52 \times 10^{-3} \text{ nm} = 1.52 \text{ pm}.$

EVALUATE: The upper level in the transition is lowered in energy so the transition energy is decreased. A smaller ΔE means a larger λ; the magnetic field increases the wavelength. The fractional shift in wavelength, $\Delta \lambda / \lambda$ is small, only 1.2×10^{-5}.

41.63. **IDENTIFY and SET UP:** The energy due to the interaction of the electron with the magnetic field is $U = -\mu_z B$. In a transition from the $m_s = -\tfrac{1}{2}$ state to the $m_s = \tfrac{1}{2}$ state, $\Delta U = 2\mu_z B$. This energy difference is the energy of the absorbed photon. The energy of the photon is $E = hf$, and $f\lambda = c$. For an electron, $S_z = \hbar/2$.

EXECUTE: **(a)** First find the frequency for each wavelength in the table with the problem. For example for the first column, $f = c/\lambda = (2.998 \times 10^8 \text{ m/s})/(0.0214 \text{ m}) = 1.40 \times 10^{10} \text{ Hz}$. Doing this for all the wavelengths gives the values in the following table. Figure 41.63 shows the graph of f versus B for this data. The slope of the best-fit straight line is 2.84×10^{10} Hz/T.

B (T)	0.51	0.74	1.03	1.52	2.02	2.48	2.97
$f(10^{10}$	1.401	2.097	2.802	4.199	5.604	7.005	8.398

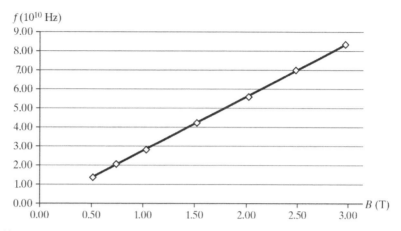

Figure 41.63

(b) The photon energy is $E = hf$, and that is the energy difference between the two levels. So

$E = \Delta U = 2\mu_z B = hf$, which gives $f = \left(\dfrac{2\mu_z}{h}\right) B$. The graph of f versus B should be a straight line

having slope equal to $2\mu_z / h$. Therefore the magnitude of the spin magnetic moment is

$\mu_z = \frac{1}{2} h(\text{slope}) = \frac{1}{2} (6.626 \times 10^{-34} \text{ J·s}) (2.84 \times 10^{10} \text{ Hz/T}) = 9.41 \times 10^{-24} \text{ J/T}.$

(c) Using $\gamma = \dfrac{|\mu_z|}{|S_z|} = \dfrac{|\mu_z|}{\hbar/2}$ gives $\gamma = \dfrac{2(9.41 \times 10^{-24} \text{ J/T})}{1.055 \times 10^{-34} \text{ J·s}} = 1.784 \times 10^{11} \text{ Hz/T},$ which rounds to

1.78×10^{11} Hz/T. This gives $\dfrac{\gamma}{e/2m} = \dfrac{2\gamma m}{e} = \dfrac{2(1.784 \times 10^{11} \text{ Hz/T})(9.11 \times 10^{-31} \text{ kg})}{1.602 \times 10^{-19} \text{ C}} = 2.03.$

EVALUATE: Our result of 2.03 for the gyromagnetic ratio for electron spin is in good agreement with the currently accepted value of 2.00232.

41.65. **IDENTIFY:** The inner electrons shield much of the nuclear charge from the outer electrons, and this shielding affects the energy levels compared to the hydrogen levels. The atom behaves like hydrogen with an effective charge in the nucleus.

SET UP: The ionization energies are $E_n = \dfrac{Z_{\text{eff}}^2}{n^2} (13.6 \text{ eV}).$

EXECUTE: **(a)** Make the conversions requested using the following conversion factor.

$E(\text{eV/atom}) = E(\text{kJ/mol})\left(\dfrac{1 \text{ mol}}{6.02214 \times 10^{23} \text{ atoms}}\right)\left(\dfrac{1000 \text{ J}}{1 \text{ kJ}}\right)\left(\dfrac{1 \text{ eV}}{1.602 \times 10^{-19} \text{ J}}\right) = 0.010364 E(\text{kJ/mol}).$

Li: $E = (0.010364)(520.2 \text{ kJ/mol}) = 5.391$ eV
Na: $E = (0.010364)(495.8 \text{ kJ/mol}) = 5.139$ eV
K: $E = (0.010364)(418.8 \text{ kJ/mol}) = 4.341$ eV
Rb: $E = (0.010364)(403.0 \text{ kJ/mol}) = 4.177$ eV
Cs: $E = (0.010364)(375.7 \text{ kJ/mol}) = 3.894$ eV
Fr: $E = (0.010364)(380 \text{ kJ/mol}) = 3.9$ eV.

(b) From the periodic chart in the Appendix, we get the following information.
Li: $Z = 3, n = 2$
Na: $Z = 11, n = 3$
K: $Z = 19, n = 4$
Rb: $Z = 37, n = 5$
Cs: $Z = 55, n = 6$
Fr: $Z = 87, n = 7$.

(c) Use $E_n = \dfrac{Z_{\text{eff}}^2}{n^2}(13.6 \text{ eV})$ and the values of n in part (b) to calculate Z_{eff}. For example, for Li we have

$$5.391 \text{ eV} = \frac{Z_{\text{eff}}^2}{2^2}(13.6 \text{ eV}) \rightarrow Z_{\text{eff}} = 1.26, \text{ and for Na we have}$$

$$5.139 \text{ eV} = \frac{Z_{\text{eff}}^2}{3^2}(13.6 \text{ eV}) \rightarrow Z_{\text{eff}} = 1.84. \text{ Doing this for the other atoms gives}$$

Li: $Z_{\text{eff}} = 1.26$
Na: $Z_{\text{eff}} = 1.84$
K: $Z_{\text{eff}} = 2.26$
Rb: $Z_{\text{eff}} = 2.77$
Cs: $Z_{\text{eff}} = 3.21$
Fr: $Z_{\text{eff}} = 3.8$.

EVALUATE: **(d)** We can see that Z_{eff} increases as Z increases. The outer (valence) electron has increasing probability density within the inner shells as Z increases, and therefore it "sees" more of the nuclear charge.

41.67. **IDENTIFY:** This problem deals with the entanglement of three spin-3/2 particles.

SET UP: $s_z = m_z \hbar, m_z = \pm 1/2, \pm 3/2$.

EXECUTE: **(a)** These particles have spin 3/2, so their spin magnitude S is

$$S = \sqrt{\frac{3}{2}\left(\frac{3}{2}+1\right)}\,\hbar = \sqrt{15/4}\,\hbar.$$

(b) No particles can have identical states. We follow the reasoning of Section 41.8 in the textbook. For simplicity call the states $A = 1/2, B = -1/2, C = -3/2$. The possible combinations of ABC are $ABC, ACB, BAC, BCA, CAB,$ and CBA. In terms of the wave functions, the normalized wave functions are

$$\psi = \frac{1}{\sqrt{6}}\left(\psi_{+1/2}\psi_{-1/2}\psi_{-3/2} - \psi_{+1/2}\psi_{-3/2}\psi_{-1/2} + \psi_{-1/2}\psi_{+1/2}\psi_{-3/2} - \psi_{+1/2}\psi_{-3/2}\psi_{+1/2} + \psi_{-3/2}\psi_{+1/2}\psi_{-1/2} \right.$$

$$\left. - \psi_{-3/2}\psi_{-1/2}\psi_{+1/2}\right).$$

The factor of $1/\sqrt{6}$ is present for normalization because the probability is proportional to the square of the wave function.

(c) Particle A has $m_z = +1/2$, but the other two could have $m_z = -3/2$ or $-1/2$. So the wave function is now $\psi = \dfrac{1}{\sqrt{2}}\left(\psi_{+1/2}\psi_{-3/2}\psi_{-1/2} - \psi_{+1/2}\psi_{+1/2} + \psi_{-3/2}\right)$.

(d) Particle A has $m_z = +1/2$, so particle B cannot have $m_z = +1/2$. But it could have $m_z = -1/2$ or $-3/2$. None of them can have $m_z = +3/2$. So the probabilities are 1/2, 0, 1/2, 0.

(e) Now we know that particle C has $m_z = -3/2$, so the wave functions are $\psi = \psi_{+1/2}\,\psi_{-1/2}\psi_{-3/2}$.

(f) We know that $m_z = -1/2$ for particle B, so the probabilities are 0, 1, 0, 0.

EVALUATE: We could not draw all these conclusions if the particles were not entangled.

STUDY GUIDE FOR MOLECULES AND CONDENSED MATTER

Summary

In this chapter, we'll extend our application of quantum mechanics from molecules and larger structures of atoms. We'll investigate chemical behavior, molecular bonds, and the large-scale assembly of atoms into crystalline solids. Our quantum-mechanical foundation will allow us to examine semiconductors and superconductors, two types of materials having profound effects on science and society today.

Objectives

After studying this chapter, you'll understand

- Forms of bonds that hold atoms together.

- The rotational and vibrational dynamics of molecules.

- How atoms form crystalline structures.

- How to use the energy-band structure to explain electrical properties of solids.

- The basis of semiconducting materials and how semiconductor devices operate.

- The basis of superconducting materials.

From Chapter 42 of Student's Study Guide to accompany *University Physics with Modern Physics, Volume 3,* Fifteenth Edition.
Hugh D. Young and Roger A. Freedman. Copyright © 2020 by Pearson Education, Inc. All rights reserved.

Concepts and Equations

Term	Description
Molecular Bonds and Molecular Spectra	Molecules bind through ionic, covalent, van der Waals, and hydrogen bonds. For a diatomic molecule, the rotational energy levels are given by $$E_l = l(l+1)\frac{\hbar^2}{2I} \qquad l = 0, 1, 2, \ldots$$ $$I = m_r r_0^2, \qquad m_r = \frac{m_1 m_2}{m_1 + m_2},$$ where I is the moment of inertia of the molecule, m_r is the reduced mass of the molecules, and r_0 is the distance between the atoms. The vibrational levels are given by $$E_n = \left(n + \frac{1}{2}\right)\hbar\omega = \left(n + \frac{1}{2}\right)\hbar\sqrt{\frac{k'}{m_r}} \qquad n = 0, 1, 2, \ldots.$$
Free-electron Model of Metals	In the free-electron model of metals, electrons are treated as free particles in a conductor. The density of states is given by $$g(E) = \frac{(2m)^{3/2} V}{2\pi^2 \hbar^3} E^{1/2}.$$ The probability that an energy state E is occupied is given by the Fermi–Dirac distribution, $$f(E) = \frac{1}{e^{(E - E_F)/kT} + 1},$$ where E_F is the Fermi energy.
Semiconductors	A semiconductor is a material with electrical resistivity intermediate between that of a good conductor and a good insulator. In n-type semiconductors, the conductivity is due to the motion of electrons. In p-type semiconductors, the conductivity is due to the motion of holes, or vacancies of electrons. Semiconductors have energy gaps of about 1 eV between their valence and conduction bands.
Semiconducting Devices	A diode is made of two semiconductors, one p-type and one n-type, that can behave much like a switch, conducting above a threshold voltage and insulating below. A transistor is made of two p–n junctions. The current–voltage relationship for an ideal p–n junction is given by $$I = I_S(e^{eV/kT} - 1).$$

Key Concept 1: In an ionic bond, the electric potential energy of the two ions (one positive and one negative) is negative and roughly the same as if we consider the two ions as point charges.

Key Concept 2: Because angular momentum is quantized (quantum number l), the rotational kinetic energy of a diatomic molecule is quantized as well. This energy is proportional to $l(l+1)$ and inversely proportional to the moment of inertia of the molecule for an axis through the molecule's center of mass. Transitions between rotational states that involve photon emission or absorption are allowed only if l changes by 1.

Key Concept 3: In addition to rotational motion around its center of mass (quantum number l), a diatomic molecule can have vibrational motion of its atoms. The vibrational energy levels (quantum number n) are evenly spaced in energy. Transitions between vibrational states that involve photon emission or absorption are allowed only if n and l both change by 1.

Key Concept 4: An ionic crystal is stable because the net electric potential energy of its ions is less than if the crystal were taken apart and the ions were moved far from each other.

Key Concept 5: The electron energy levels in a semiconductor form a valence band that is essentially filled, plus a conduction band that is empty and separated from the valence band by a small energy gap. The conductivity of a semiconductor can be increased by using photons to excite electrons from the valence band into the conduction band. To do this, the photon energy must be at least as great as the band gap; lower-energy photons will not be absorbed.

Key Concept 6: In the free-electron model of a metal, the Fermi–Dirac distribution gives the probability that a state of a given energy E is occupied at a temperature T. The probability is greater than ½ for energies less than the Fermi energy E_F and less than ½ for energies greater than E_F.

Key Concept 7: The Fermi energy in the free-electron model of a metal depends on the electron concentration (number of electrons per unit volume). In a typical metal at room temperature T, the Fermi energy is much greater than kT.

Key Concept 8: The exclusion principle, not the effects of temperature, is the dominant reason for the distribution of electron energies in a metal at room temperature.

Key Concept 9: The properties of the Fermi–Dirac distribution explain why the number of electrons in the conduction band of a semiconductor is sensitive to both the temperature and the size of the band gap.

Conceptual Questions

1: Types of bonds

What kind of chemical bonds hold the following objects together: (a) NaCl molecules, (b) N_2 molecules, and (c) copper atoms in a wire.

IDENTIFY, SET UP, AND EXECUTE (a) NaCl molecules are made of oppositely charged atoms, so ionic bonding holds the molecules together. The sodium atom gives its one $3s$ electron to the chlorine atom, filling a vacancy in the chlorine's $3p$ subshell.
(b) Since two nitrogen atoms form a nitrogen molecule, the molecule is made of two similarly charged atoms and cannot be bound by an ionic bond. The nitrogen molecule is bound by a covalent bond.
(c) The copper atoms in a wire are bound by metallic bonds, since the copper is arranged in a crystalline structure in the wire.

EVALUATE This small sample of problems involving bonds illustrates the variations on chemical bonding in molecules and solids.

2: Covalent bonds without electron spin

How many electrons would participate in a covalent bond if electrons had no spin?

IDENTIFY, SET UP, AND EXECUTE If electrons had no spin, only one electron could occupy a particular space, due to the exclusion principle. The usual covalent bond has two electrons sharing the same spatial state, but not the same spin state, in order to satisfy the exclusion principle. Without spin, only one electron could participate in a covalent bond.

EVALUATE Spin plays an important role in atomic and molecular physics, as we see here in its role in covalent bonds. This exercise should elucidate the role spin plays in molecular binding. True spinless particles exist, but they do not obey the exclusion principle.

From Chapter 42 of Student's Study Guide to accompany *University Physics with Modern Physics, Volume 3,* Fifteenth Edition.
Hugh D. Young and Roger A. Freedman. Copyright © 2020 by Pearson Education, Inc. All rights reserved.

Problems

1: Predicting states

Light of wavelength 5.0 μm strikes and is absorbed by a molecule. Is this process most likely to alter the rotational, vibrational, or atomic energy levels of the molecule? If the light had a wavelength of 3.7 mm, which energy levels would most likely be affected?

IDENTIFY AND SET UP The energy difference for atomic energy levels is generally several eVs, for vibrational levels is generally several 0.1 eVs, and for rotational levels is generally several 0.001 eVs. We'll calculate the energy of the light to determine which type of transition it may affect.

EXECUTE The energy of a photon is

$$E = \frac{hc}{\lambda}.$$

Since we are calculating electron-volts, it is convenient to use $h = 4.136 \times 10^{-15}$ eVs. The energy of the 5.0 - μm light is then

$$E = \frac{hc}{\lambda} = \frac{(4.136 \times 10^{-15} \text{ eV} \cdot \text{s})(3.00 \times 10^8 \text{ m/s})}{5.0 \times 10^{-6} \text{ m}} = 0.25 \text{ eV}.$$

0.25 eV corresponds to the energy difference of a vibrational transition. For the 3.7-mm light, the energy is

$$E = \frac{hc}{\lambda} = \frac{(4.136 \times 10^{-15} \text{ eV} \cdot \text{s})(3.00 \times 10^8 \text{ m/s})}{3.7 \times 10^{-3} \text{ m}} = 0.0003 \text{ eV}.$$

0.0003 eV corresponds to the energy difference of a rotational transition. The 5.0 μm light should affect the vibrational level, and the 3.7-mm light should affect the rotational level.

EVALUATE These different ranges of transition energies indicate the likelihood of causing a molecule to rotate, vibrate, or change its atomic state. The situation corresponds to a similar one in the classical world, where it is easy to rotate an object, harder to induce a vibration, and much harder to have it change state.

Practice Problem: If the light had a wavelength of 140 nm, which energy levels would most likely be affected? *Answer:* Light of this wavelength has an energy of 8.9 eV, corresponding to an atomic transition.

Extra Practice: What wavelength of light will create an energy difference of 1 eV? *Answer:* 1240 nm

2: A diatomic molecule

The D_2 molecule is made up of two deuterons (a proton plus a neutron) and two electrons. The spacing between the nuclei in the molecule is approximately 7.5×10^{-11} m. Calculate the moment of inertia of the D_2 molecule about its center of mass and the rotational energy of its ground state and first two excited states.

IDENTIFY AND SET UP We'll use the formulae for the moment of inertia and rotational energy to solve the problem. We'll ignore the masses of the electrons, as they are negligible compared with the mass of the nuclei. We also will take the mass of the neutron to be that of the proton, as they have similar masses.

EXECUTE The moment of inertia is

$$I = \sum m_i r_i^2.$$

There are four masses, each located at a point half the interatomic distance from the center of mass. This gives

$$I = 2m_H \left(\frac{d}{2}\right)^2 + 2m_H \left(\frac{d}{2}\right)^2 = m_H d^2$$

$$= (1.67 \times 10^{-27} \text{ kg})(7.5 \times 10^{-11} \text{ m})^2 = 9.39 \times 10^{-48} \text{ kg m}^2.$$

The energy levels for rotational states are given by

$$E = l(l+1)\frac{(\hbar)^2}{2I}$$

$$= l(l+1)(3.70 \times 10^{-3} \text{ eV}).$$

The ground-state energy has $l = 0$, so $E_0 = 0$; the first excited-state energy has $l = 1$, so $E_1 = 7.40 \times 10^{-3}$ eV; and the second excited-state energy has $l = 2$, so $E_2 = 2.22 \times 10^{-2}$ eV.

KEY CONCEPT 2 **EVALUATE** What frequency of light will be emitted when the first excited state decays to the ground state?

Practice Problem: What is the energy of the third excited state? *Answer:* 4.44×10^{-2} eV

Extra Practice: What frequency of light is emitted when the third excited state decays to the first excited state? *Answer:* 1.71×10^{-2} eV

3: Fermi energy of a particle in a box

Suppose the energy values for a particle in a box are given by

$$E_n = E_0 n^2,$$

where E_0 is a constant and n is the quantum number of the state. (a) If there are 50 electrons in such states, find the Fermi energy at a temperature of 0 K. (b) What is the ratio of the average energy of the electrons to the Fermi energy?

IDENTIFY AND SET UP The Fermi energy is the energy of the last filled state, so we'll find the energy of the highest filled state to solve part (a). To solve part (b), we'll add up the energies of all the states and take the ratio.

EXECUTE (a) The 50 electrons will occupy 25 states, since 2 electrons fill each state. The energy of the 25th state is

$$E_{F0} = E_0(25)^2 = 625E_0.$$

(b) The total energy is found by summing the contributions from each state, given by

$$E_T = 2E_0[1 + 4 + 9 + \cdots] = 2E_0 \sum_{n=1}^{25} n^2.$$

We can solve this equation by summing up the 25 terms, or we can use a summation rule. The sum over N squares is given by

$$\sum_{n=1}^{N} n^2 = \frac{N(N+1)(2N+1)}{6}.$$

Evaluating this sum for $N = 25$, we have

$$E_T = 2E_0 \sum_{n=1}^{25} n^2 = 2E_0 \frac{25(25+1)(2(25)+1)}{6} = 11{,}050E_0.$$

The average energy is the total energy divided by 50:

$$E_{avg} = \frac{E_T}{50} = \frac{11{,}050}{50} E_0 = 221E_0 = \left(\frac{221}{625}\right) E_{F0}.$$

The ratio of the average energy per electron to the Fermi energy is 0.354.

EVALUATE What does the ratio of the average energy per electron to the Fermi energy tell us about the system? Since the ratio is about 1/3, it indicates that many of the electrons are in energy states having less than half the Fermi energy.

Practice Problem: What is the Fermi energy at 0 K if there are 40 electrons? *Answer:* 400 E_0

Extra Practice: What is the ratio of the average energy of the electrons to the Fermi energy? *Answer:* 0.358

4: Fermi–Dirac statistics

Given the Fermi–Dirac probability distribution and the fact that sodium has a Fermi energy of 3.15 eV, find the ratio of the width ΔE to E_F at 273 K, where $\Delta E = E(0.2) - E(0.8)$. The quantity $E(0.2)$ is the energy for which the occupation probability is 0.2.

IDENTIFY AND SET UP Use the Fermi–Dirac probability function to solve the problem.

EXECUTE The Fermi–Dirac probability function $f(E)$ is given by

$$f(E) = \frac{1}{e^{(E-E_F)/kT} + 1}.$$

Rearranging terms to solve for E gives

$$E = E_F + kT \ln\left(\frac{1-f(E)}{f(E)}\right).$$

Solving for $E(0.2)$ and $E(0.8)$ yields

$$E(0.2) = E_F + kT \ln\left(\frac{1-0.2}{0.2}\right) = E_F + kT \ln 4$$

$$E(0.8) = E_F + kT \ln\left(\frac{1-0.8}{0.8}\right) = E_F - kT \ln 4.$$

Forming the ratio requested in the problem gives

$$\frac{\Delta E}{E_F} = \frac{2kT \ln 4}{E_F} = \frac{2(8.617 \times 10^{-5} \text{ eV/K})(273 \text{ K}) \ln 4}{(3.15 \text{ eV})} = 0.02.$$

KEY CONCEPT 6 **EVALUATE** What does this ratio tell us about the system? Near room temperature, changing the energy by 2% of the Fermi energy changes the occupation probability for sodium from 80% to 20%. This is a very dramatic change over a small distance (i.e., the Fermi distribution function is very sharp).

Try It Yourself!

1: Rotational transitions

(a) If a molecule with moment of inertia I is induced to make a pure rotational transition from a state L to a state $L+1$, what frequency of radiation is needed? (b) If the same molecule makes a transition from the state L to the state $L-1$, what frequency of radiation is emitted?

Solution Checkpoints

IDENTIFY AND SET UP Use rotational energy levels to solve the problem.

EXECUTE (a) The rotational energy levels are given by

$$E_l = l(l+1)\frac{\hbar^2}{2I}.$$

To go from state L to state $L+1$, the energy hf supplied must be equal to

$$hf = E_{L+1} - E_L$$

$$= 2(L+1)\frac{\hbar^2}{2I}.$$

(b) To go from state L to state $L-1$, the energy hf released is equal to

$$hf' = E_L - E_{L-1}$$

$$= 2L\frac{\hbar^2}{2I}.$$

KEY CONCEPT 2 **EVALUATE** How do the spacing between different possible values of f and f' compare?

2: Promoting to the conduction band

Electromagnetic radiation can promote an electron from the top of a nearly complete valence band to the bottom of an unfilled conduction band. The lowest frequency for which this is possible is $f_m = 2.75 \times 10^{14}$ Hz for silicon and $f_m = 1.79 \times 10^{14}$ Hz for germanium, both at room temperature. Calculate the energy gap between the valence and conduction bands for the two materials.

Solution Checkpoints

IDENTIFY AND SET UP What equation relates energy and frequency for electromagnetic radiation?

EXECUTE The energy gap is found from the frequency by the relationship

$$\Delta E = hf_m = E_{gap}.$$

From Chapter 42 of Student's Study Guide to accompany *University Physics with Modern Physics, Volume 3,* Fifteenth Edition.
Hugh D. Young and Roger A. Freedman. Copyright © 2020 by Pearson Education, Inc. All rights reserved.

Evaluating this equation gives an energy gap of 1.13 eV for silicon and an energy gap of 0.736 eV for germanium.

KEY CONCEPT 5 **EVALUATE** What practical purpose can this material carry out? Could it be used to detect light? How?

Practice Problem: What maximum wavelength of light can promote an electron in a material with an energy gap of 1.5 eV? *Answer:* 827 nm

Extra Practice: What is the smallest gap for which green light (495–570 nm) can promote an electron? *Answer:* 2.18 eV

Key Example Variation Problems

Solutions to these problems are in Chapter 42 of the Student's Solutions Manual.

Be sure to review EXAMPLES 42.2 and 42.3 (Section 42.2) before attempting these problems.

VP42.3.1 The two nuclei in the nitric oxide (NO) molecule are 0.1154 nm apart. The mass of the most common nitrogen atom is 2.326×10^{-26} kg, and the mass of the most common oxygen atom is 2.656×10^{-26} kg. Find (a) the reduced mass of the NO molecule, (b) the moment of inertia of the NO molecule, and (c) the energies, in meV, of the lowest three rotational energy levels of NO.

VP42.3.2 The $l = 2$ rotational level of the hydrogen chloride (HCl) molecule has energy 7.90 meV. The mass of the most common hydrogen atom is 1.674×10^{-27} kg, and the mass of the most common chlorine atom is 5.807×10^{-26} kg. Find (a) the moment of inertia of the HCl molecule, (b) the reduced mass of the HCl molecule, and (c) the distance between the H and Cl nuclei.

VP42.3.3 The mass of the most common silicon atom is 4.646×10^{-26} kg, and the mass of the most common oxygen atom is 2.656×10^{-26} kg. When a molecule of silicon monoxide (SiO) makes a transition between the $l = 1$ and $l = 0$ rotational levels, it emits a photon of wavelength 6.882 mm. Find (a) the moment of inertia of the SiO molecule, (b) the reduced mass of the SiO molecule, and (c) the distance between the Si and O nuclei.

VP42.3.4 A CO molecule is initially in the $n = 2$ vibrational level. If this molecule loses both vibrational and rotational energy and emits a photon, what are the photon wavelength and frequency if the initial angular momentum quantum number is (a) $l = 3$ and (b) $l = 2$?

Be sure to review EXAMPLES 42.6 and 42.7 (Section 42.5) before attempting these problems.

VP42.7.1 For free electrons in a solid with Fermi energy E_F and temperature T, find the energy for which the probability that a state at that energy is occupied is (a) 0.33 and (b) 0.90.

VP42.7.2 The Boltzmann constant is $k = 8.617 \times 10^{-5}$ eV/K. For a metallic solid at room temperature (293 K), what is the probability that an electron state is occupied if its energy is (a) 0.0250 eV below the Fermi level, (b) 0.0400 eV above the Fermi level, and (c) 0.100 eV above the Fermi level?

VP42.7.3 Silver contains 5.8×10^{28} free electrons per cubic meter. At absolute zero, what are (a) the Fermi energy (in J and eV) of silver, (b) the speed of an electron with this energy, and (c) the density of states (in states/J and states/eV) at the Fermi energy for a block of silver of volume 1.0 cm³?

VP42.7.4 If kT is small compared to the Fermi energy at absolute zero, the Fermi energy at temperature T is essentially the same as at absolute zero. Use the results of the previous problem to find the electron energy, in eV, for which the probability that an electron state is occupied is (a) 0.92 and (b) 1.0×10^{-4} for silver at room temperature (293 K).

Be sure to review EXAMPLE 42.9 (Section 42.6) before attempting these problems.

VP42.9.1 For a certain semiconductor, the Fermi energy is in the middle of its band gap. If the temperature of the semiconductor is 285 K, find the probability that a state at the bottom of the conduction band is occupied if the band gap is (a) 0.500 eV and (b) 1.50 eV.

VP42.9.2 Find the ratio of the probability that a state at the bottom of the conduction band of a semiconductor is occupied at 315 K to the probability at 295 K if the band gap is (a) 0.400 eV and (b) 0.800 eV. Assume that the Fermi energy is in the middle of the band gap.

VP42.9.3 Consider a material whose Fermi energy is in the middle of its band gap. What band-gap width provides a probability of 0.00190 that a state at the bottom of the conduction band is occupied if the temperature of the material is (a) 305 K and (b) 325 K?

VP42.9.4 In a certain semiconductor, the Fermi energy lies above the top of the valence band by an amount equal to 34 of the band gap. Find the probability that a state at the bottom of the conduction band is occupied if the band gap is 0.300 eV and the temperature is (a) 275 K and (b) 325 K.

STUDENT'S SOLUTIONS MANUAL FOR MOLECULES AND
CONDENSED MATTER

VP42.3.1. **IDENTIFY:** This problem involves the rotational energy of a molecule.
SET UP and EXECUTE: **(a)** We want the reduced mass. Use the given masses in the equation

$$m_r = \frac{m_1 m_2}{m_1 + m_2}.$$

The result is $m_r = 1.240 \times 10^{-26}$ kg.
(b) We want the moment of inertia. Use $r_0 = 0.1154$ nm and the result of part (a). The result is
$I = m_r r_0^2 = 1.651 \times 10^{-46}$ kg·m^2.
(c) We want the energies of the rotational states $l = 0, 1, 2$.

$$E_l = l(l+1)\frac{\hbar^2}{2I}$$

$$E_0 = 0$$

$$E_1 = 1(1+1)\frac{\hbar^2}{2I} = 4.203 \times 10^{-4} \text{ eV} = 0.4203 \text{ meV}.$$

$$E_2 = 2(2+1)\frac{\hbar^2}{2I} = 1.261 \times 10^{-3} \text{ eV} = 1.261 \text{ meV}.$$

EVALUATE: Note that molecular rotational energies are of the order of *milli*electron-volts, whereas energy states for electrons in atoms are around a few eV.

VP42.3.2. **IDENTIFY:** We are dealing with the rotational energy levels of a molecule.
SET UP: The following equations apply:

$$E_l = l(l+1)\frac{\hbar^2}{2I}, \quad m_r = \frac{m_1 m_2}{m_1 + m_2}, \quad I = m_r r_0^2.$$

EXECUTE: **(a)** We want the moment of inertia. Solve the E_l equation for I for the $l = 2$ state.

$$I = \frac{3\hbar^2}{E_2} = \frac{3\hbar^2}{7.90 \text{ meV}} = 2.64 \times 10^{-47} \text{ kg·m}^2.$$

(b) We want the reduced mass. Use the two given masses in the equation for m_r, giving

$$m_r = 1.627 \times 10^{-27} \text{ kg}.$$

(c) We want the distance r_0 between the two nuclei. Solve $I = m_r r_0^2$ for r_0 and use the results of parts (a) and (b). This gives

$$r_0 = \sqrt{I/m_r} = 0.127 \text{ nm}.$$

EVALUATE: Note that the chlorine has only a small effect on the reduced mass which is essentially the mass of the hydrogen atom.

VP42.3.3. **IDENTIFY:** This problem is about the rotational energy levels of the SiO molecule.
SET UP: The following equations apply:

$$E_l = l(l+1)\frac{\hbar^2}{2I}, \quad m_r = \frac{m_1 m_2}{m_1 + m_2}, \quad I = m_r r_0^2.$$

EXECUTE: **(a)** We want I. The energy of the emitted photon is equal to the energy difference between the $l = 0$ and $l = 1$ rotational levels. The photon energy is $E = hc/\lambda$. Therefore

$$E_1 - E_0 = 1(1+1)\frac{\hbar^2}{2I} - 0 = \frac{\hbar^2}{I}$$

Solving for I gives

$$I = \frac{\hbar\lambda}{2\pi c} = \frac{\hbar(6.882\text{ mm})}{2\pi c} = 3.853\times10^{-46} \text{ kg} \cdot \text{m}^2.$$

(b) We want the reduced mass. Use the given masses.

$$m_r = \frac{m_1 m_2}{m_1 + m_2} = 1.690\times10^{-26} \text{ kg}.$$

(c) We want r_0. Solve $I = m_r r_0^2$ for r_0 and use the answers from parts (a) and (b), which gives r_0 = 0.1510 nm.
EVALUATE: The result in part (c) is quite reasonable for atomic distances in molecules.

VP42.3.4. **IDENTIFY:** This problem involves the vibrational and rotational energy levels in the CO molecule.
SET UP: When the molecule makes a transition to a lower energy state, the energy E_{ph} of the emitted photon is equal to the *magnitude* of the energy difference between the two states. Using Eq. (42.9) this photon energy is

$$E_{ph} = l_1(l_1 +1)\frac{\hbar^2}{2I} + \left(n_1 + \frac{1}{2}\right)\hbar\omega - \left[l_2(l_2 +1)\frac{\hbar^2}{2I} + \left(n_2 + \frac{1}{2}\right)\hbar\omega\right]$$

$$E_{ph} = [l_1(l_1 +1) - l_2(l_2 +1)]\frac{\hbar^2}{2I} + (n_1 - n_2)\hbar\omega$$

$n_1 - n_2 = +1$. Using this fact and some results from Example 42.3, we have

$$(n_1 - n_2)\hbar\omega = 0.2690 \text{ eV}, \quad \frac{\hbar^2}{2I} = 0.2395 \text{ meV}$$

The photon energy is therefore given by

$$E_{ph} = [l_1(l_1 +1) - l_2(l_2 +1)](0.2395 \text{ meV}) + 0.2690 \text{ eV}$$

EXECUTE: **(a)** Initial state is $l_1 = 3$. $\Delta l = -1$, so $l_2 = 2$.
$3 \rightarrow 2$ transition:
$$E_{ph} = [3(3+1) - 2(2+1)](0.2395 \text{ meV}) + 0.2690 \text{ eV} = 1.4370 \text{ meV} + 0.2690 \text{ eV} = 0.270437 \text{ eV}.$$

$$f = E_{ph}/h = (0.270437 \text{ eV})/h = 6.535 \times 10^{13} \text{ Hz}.$$
$$\lambda = c/f = 4.585 \ \mu\text{m}.$$

(b) Initial state is $l_1 = 2$. $\Delta l = -1$, so $l_2 = 1$.
$2 \rightarrow 1$ transition:
$$E_{ph} = [2(2+1) - 1(1+1)](0.2395 \text{ meV}) + 0.2690 \text{ eV} = 0.9580 \text{ meV} + 0.2690 \text{ eV} - 0.269958 \text{ eV}$$

$$f = E_{ph}/h = (0.269958 \text{ eV})/h = 6.538 \times 10^{13} \text{ Hz}.$$
$$\lambda = c/f = 4.593 \ \mu\text{m}.$$

EVALUATE: The wavelengths are very close together because the energy differences between states are very small since they are due only to *rotational* transitions which are low energy.

VP42.7.1. **IDENTIFY:** This problem involves the Fermi-Dirac distribution.
SET UP: Eq. (42.16) gives the probability $f(E)$ that a given energy state is occupied. The target variable is the energy for each probability. Solve for E in each case.
EXECUTE: **(a)** $f(E) = 0.33$. Solve for E.

$$0.33 = \frac{1}{e^{(E-E_F)/kT} + 1}$$

Separate the exponential and use logarithms, giving
$$(E - E_F)/kT = \ln 2.02$$
$$E = E_F + 0.71kT.$$

(b) $f(E) = 0.90$. Follow the same procedure using 0.90 instead of 0.33, giving $E = E_F - 2.2kT$.
EVALUATE: If $E = E_F$, $f(E) = 0.50$, for $f(E) > 0.50$, $E < E_F$, and for $f(E) < 0.50$, $E > E_F$. Our results are consistent with these conditions.

VP42.7.2. **IDENTIFY:** This problem involves the Fermi-Dirac distribution.
SET UP: Eq. (42.16) gives the probability $f(E)$ that a given energy state is occupied. The target variable is the probability for each specified energy.
EXECUTE: **(a)** $E = E_F - 0.0250$ eV. Using Eq. (42.16) gives

$$f(E) = \frac{1}{e^{(E_F - 0.0250\ \text{eV} - E_F)/kT} + 1}$$

The exponent of e is $(0.0250\ \text{eV})/kT = 0.9902$, so $f(E) = 1/(e^{-0.9902} + 1) = 0.729$.
(b) $E = E_F + 0.0400$ eV. Using Eq. (42.16) and following the same procedure gives $f(E) = 1/(e^{1.5843} + 1) = 0.170$.
(c) $E = E_F + 0.100$ eV. Using the same procedure gives $f(E) = 0.0187$.
EVALUATE: The probability decreases as E increases above the Fermi energy.

VP42.7.3. **IDENTIFY:** This problem involves the Fermi energy and the density of states.
SET UP and EXECUTE: **(a)** We want the Fermi energy at absolute zero. Using Eq. (42.19) with N/V given in the problem, we get

$$E_{F0} = \frac{3^{2/3}\pi^{4/3}\hbar^2}{2m}\left(\frac{N}{V}\right)^{2/3} = 5.5\ \text{eV}.$$

(b) The speed is the target variable. The kinetic energy K is equal to the Fermi energy. Solve $K = E_F$ for v and use the result from part (a).

$$v = \sqrt{\frac{2E_F}{m}} = 1.4\times10^6\ \text{m/s}.$$

(c) The target variable is the density of states. Use Eq. (42.15) with $V = 1.0\ \text{cm}^3$ and $E = 5.5$ eV from part (a). The result is

$$g(E) = \frac{(2m)^{3/2}V}{2\pi^2\hbar^3}E^{1/2} = 9.9\times10^{40}\ \text{states/J} = 1.6\times10^{22}\ \text{states/eV}.$$

EVALUATE: The electron speeds are not great enough to require the use of special relativity.

VP42.7.4. **IDENTIFY:** This problem deals with the Fermi energy.
SET UP: From the previous problem we have $E_{F0} = 5.5$ eV. The target variable is the electron energy for the given probabilities.
EXECUTE: **(a)** $f(E) = 0.92$. Solve Eq. (42.16) for E using logarithms.

$$0.92 = \frac{1}{e^{(E-E_F)/kT} + 1}$$
$$E - E_F = kT\ln(0.08696)$$
$$E = 5.4\ \text{eV}.$$

(b) $f(E) = 1.0 \times 10^{-4}$. Use the same procedure as in part (a) except $f(E) = 0.00010$, which gives $E = 5.7$ eV.

EVALUATE: Our results are consistent with problem VP42.7.1. When $E > E_F$, $f(E) < 0.50$, and when $E < E_F$, $f(E) > 0.50$.

VP42.9.1. **IDENTIFY:** This problem is about the Fermi energy and the band gap in a semiconductor.
SET UP and EXECUTE: For an electron at the bottom of the conduction band, $E = E_F + E_g/2$, so $E - E_F = E_g/2$. The target variable is the probability $f(E)$. Use Eq. (42.16) for $f(E)$.
(a) $E_g = 0.500$ eV. $(E - E_F)/kT = E_g/2kT = (0.500 \text{ eV})/2kT = 10.18$. Eq. (42.16) gives $f(E) = 1/(e^{10.18} + 1) = 3.79 \times 10^{-5}$.
(b) $E_g = 1.50$ eV. $(E - E_F)/kT = E_g/2kT = (1.50 \text{ eV})/2kT = 30.54$. Eq. (42.16) gives $f(E) = 1/(e^{30.54} + 1) = 5.46 \times 10^{-14}$.
EVALUATE: The probability decreases as the gap widens.

VP42.9.2. **IDENTIFY:** This problem is about the band gap in a semiconductor.
SET UP: $E = E_F + E_g/2$, so $E - E_F = E_g/2$. The target variable is the probability $f(E)$. Use Eq. (42.16) for $f(E)$. We want the ratio of the probabilities at different temperatures.
EXECUTE: **(a)** $E_g = 0.400$ eV. At 315 K: $E_g/2kT = (0.400 \text{ eV})/[2k(315 \text{ K})] = 7.3682$.
$f(E)_{315} = 1/(e^{7.3682} + 1) = 6.3058 \times 10^{-4}$.
At 295 K: The exponent is 7.8677, so $f(E) = 1/(e^{7.8677} + 1) = 3.8290 \times 10^{-4}$.
Now take the ratio of the probabilities using the results we just found. This gives

$$\frac{f(E)_{315}}{f(E)_{295}} = \frac{6.3058 \times 10^{-4}}{3.8290 \times 10^{-4}} = 1.65.$$

(b) $E_g = 0.800$ eV. Use the same procedure as for part (a). At 315 K: $E_g/2kT = 14.7364$ which gives $f(E) = 3.98165 \times 10^{-7}$.
At 295 K: $E_g/2kT = 15.7354$ which gives $f(E) = 1.46623 \times 10^{-7}$.

$$\frac{f(E)_{315}}{f(E)_{295}} = \frac{3.98165 \times 10^{-7}}{1.46623 \times 10^{-7}} = 2.72.$$

EVALUATE: At $T = 315$ K the probability is greater than at 295 K, which is in agreement with Example 42.9.

VP42.9.3. **IDENTIFY:** We are dealing with the energy gap in a semiconductor.
SET UP: The target variable is the width of the energy gap E_g at different temperatures. Combining $E - E_F = E_g/2$ with Eq. (42.16) gives

$$f(E) = \frac{1}{e^{E_g/2kT} + 1}$$

Solve this equation for E_g using logarithms.

$$E_g = 2kT \ln\left(1 + \frac{1}{f(E)}\right) = 2kT \ln(527.32).$$

EXECUTE: **(a)** $T = 305$ K. Using this temperature in the equation we just derived gives $E_g = 0.329$ eV.
(b) $T = 325$ K. Using the same procedure gives $E_g = 0.351$ eV.
EVALUATE: As the temperature increases, the gap width also increases.

VP42.9.4. **IDENTIFY:** We are dealing with the energy gap in a semiconductor.
SET UP: The target variable is the probability at different temperatures. Figure VP42.9.4 helps to visualize how the various quantities are related. In this case, $E = E_F + E_g/4$. Therefore $E - E_F = E_g/4 = (0.300 \text{ eV})/4 = 0.0750$ eV and $(E - E_F)/kT = (0.0750 \text{ eV})/kT = (870.37 \text{ K})/T$.

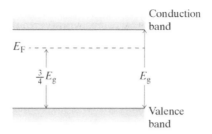

Figure VP42.9.4

EXECUTE: Use Eq. (42.16).
(a) $T = 275$ K. $(E - E_F)/kT = (870.37$ K$)/T = (870.37$ K$)/(275$ K$) = 3.1650$. Therefore

$$f(E) = \frac{1}{e^{3.1650} + 1} = 0.0405.$$

(b) $T = 325$ K. $(E - E_F)/kT = (870.37$ K$)/T = (870.37$ K$)/(325$ K$) = 2.67806$. Therefore

$$f(E) = \frac{1}{e^{2.67806} + 1} = 0.0643.$$

EVALUATE: As the temperature increases, so does the probability of finding electrons in the conduction band.

42.1. **IDENTIFY and SET UP:** $U = \frac{1}{4\pi\epsilon_0}\frac{q_1 q_2}{r}$. The binding energy of the molecule is equal to U plus the ionization energy of K minus the electron affinity of Br.

EXECUTE: **(a)** $U = -\frac{1}{4\pi\epsilon_0}\frac{e^2}{r} = -5.0$ eV.

(b) -5.0 eV $+ (4.3$ eV $- 3.5$ eV$) = -4.2$ eV.

EVALUATE: We expect the magnitude of the binding energy to be somewhat less than this estimate. At this separation the two ions don't behave exactly like point charges and U is smaller in magnitude than our estimate. The experimental value for the binding energy is -4.0 eV, which is smaller in magnitude than our estimate.

42.5. **IDENTIFY:** The energy given to the photon comes from a transition between rotational states.

SET UP: The rotational energy of a molecule is $E = l(l+1)\frac{\hbar^2}{2I}$ and the energy of the photon is $E = hc/\lambda$.

EXECUTE: Use the energy formula, the energy difference between the $l = 3$ and $l = 1$ rotational levels of the molecule is $\Delta E = \frac{\hbar^2}{2I}[3(3+1) - 1(1+1)] = \frac{5\hbar^2}{I}$. Since $\Delta E = hc/\lambda$, we get $hc/\lambda = 5\hbar^2/I$. Solving for I gives

$$I = \frac{5\hbar\lambda}{2\pi c} = \frac{5(1.055\times10^{-34} \text{ J}\cdot\text{s})(1.780 \text{ nm})}{2\pi(3.00\times10^8 \text{ m/s})} = 4.981\times10^{-52} \text{ kg}\cdot\text{m}^2.$$

Using $I = m_r r_0^2$, we can solve for r_0:

$$r_0 = \sqrt{\frac{I(m_N + m_H)}{m_N m_H}} = \sqrt{\frac{(4.981\times10^{-52} \text{ kg}\cdot\text{m}^2)(2.33\times10^{-26} \text{ kg} + 1.67\times10^{-27} \text{ kg})}{(2.33\times10^{-26} \text{ kg})(1.67\times10^{-27} \text{ kg})}} = 5.65\times10^{-13} \text{ m}.$$

EVALUATE: This separation is much smaller than the diameter of a typical atom and is not very realistic. But we are treating a *hypothetical* NH molecule.

42.7. **IDENTIFY** and **SET UP:** Set $K = E_1$ from Example 42.2. Use $K = \frac{1}{2}I\omega^2$ to solve for ω and $v = r\omega$ to solve for v.

EXECUTE: **(a)** From Example 42.2, $E_1 = 0.479$ meV $= 7.674 \times 10^{-23}$ J and $I = 1.449 \times 10^{-46}$ kg·m^2.

$K = \frac{1}{2}I\omega^2$ and $K = E$ gives $\omega = \sqrt{2E_1/I} = 1.03 \times 10^{12}$ rad/s.

(b) $v_1 = r_1\omega_1 = (0.0644 \times 10^{-9}$ m$)(1.03 \times 10^{12}$ rad/s$) = 66.3$ m/s (carbon).

$v_2 = r_2\omega_2 = (0.0484 \times 10^{-9}$ m$)(1.03 \times 10^{12}$ rad/s$) = 49.8$ m/s (oxygen).

(c) $T = 2\pi/\omega = 6.10 \times 10^{-12}$ s.

EVALUATE: Even for fast rotation rates, $v \ll c$.

42.9. **IDENTIFY** and **SET UP:** The energy of a rotational level with quantum number l is $E_l = l(l+1)\hbar^2/2I$. $I = m_r r^2$, with the reduced mass m_r given by $m_r = \dfrac{m_1 m_2}{m_1 + m_2}$. Calculate I and ΔE and then use $\Delta E = hc/\lambda$ to find λ.

EXECUTE: **(a)** $m_r = \dfrac{m_1 m_2}{m_1 + m_2} = \dfrac{m_{Li} m_H}{m_{Li} + m_H} = \dfrac{(1.17 \times 10^{-26}\ \text{kg})(1.67 \times 10^{-27}\ \text{kg})}{1.17 \times 10^{-26}\ \text{kg} + 1.67 \times 10^{-27}\ \text{kg}} = 1.461 \times 10^{-27}$ kg.

$I = m_r r^2 = (1.461 \times 10^{-27}$ kg$)(0.159 \times 10^{-9}$ m$)^2 = 3.694 \times 10^{-47}$ kg·m^2.

$l = 3 : E = 3(4)\left(\dfrac{\hbar^2}{2I}\right) = 6\left(\dfrac{\hbar^2}{I}\right)$.

$l = 4 : E = 4(5)\left(\dfrac{\hbar^2}{2I}\right) = 10\left(\dfrac{\hbar^2}{I}\right)$.

$\Delta E = E_4 - E_3 = 4\left(\dfrac{\hbar^2}{I}\right) = 4\left(\dfrac{(1.055 \times 10^{-34}\ \text{J·s})^2}{3.694 \times 10^{-47}\ \text{kg·m}^2}\right) = 1.20 \times 10^{-21}$ J $= 7.49 \times 10^{-3}$ eV.

(b) $\Delta E = hc/\lambda$, so $\lambda = \dfrac{hc}{\Delta E} = \dfrac{(4.136 \times 10^{-15}\ \text{eV})(2.998 \times 10^8\ \text{m/ s})}{7.49 \times 10^{-3}\ \text{eV}} = 166\ \mu\text{m}$.

EVALUATE: LiH has a smaller reduced mass than CO and λ is somewhat smaller here than the λ calculated for CO in Example 42.2.

42.11. **IDENTIFY:** The vibrational energy of the molecule is related to its force constant and reduced mass, while the rotational energy depends on its moment of inertia, which in turn depends on the reduced mass.

SET UP: The vibrational energy is $E_n = (n + \frac{1}{2})\hbar\omega = (n + \frac{1}{2})\hbar\sqrt{\dfrac{k'}{m_r}}$ and the rotational energy is

$E_l = l(l+1)\dfrac{\hbar^2}{2I}$.

EXECUTE: For a vibrational transition, we have $\Delta E_v = \hbar\sqrt{\dfrac{k'}{m_r}}$, so we first need to find m_r. The

energy for a rotational transition is $\Delta E_R = \dfrac{\hbar^2}{2I}[2(2+1) - 1(1+1)] = \dfrac{2\hbar^2}{I}$. Solving for I and using the

fact that $I = m_r r_0^2$, we have $m_r r_0^2 = \dfrac{2\hbar^2}{\Delta E_R}$, which gives

$$m_r = \frac{2\hbar^2}{r_0^2 \Delta E_R} = \frac{2(1.055\times10^{-34}\ \text{J}\cdot\text{s})(6.583\times10^{-16}\ \text{eV}\cdot\text{s})}{(0.8860\times10^{-9}\ \text{m})^2 (8.841\times10^{-4}\ \text{eV})} = 2.0014\times10^{-28}\ \text{kg}.$$

Now look at the vibrational transition to find the force constant.

$$\Delta E_v = \hbar\sqrt{\frac{k'}{m_r}} \Rightarrow k' = \frac{m_r(\Delta E_v)^2}{\hbar^2} = \frac{(2.0014\times10^{-28}\ \text{kg})(0.2560\ \text{eV})^2}{(6.583\times10^{-16}\ \text{eV}\cdot\text{s})^2} = 30.27\ \text{N/m}.$$

EVALUATE: This would be a rather weak spring in the laboratory.

42.15. **IDENTIFY:** The energy gap is the energy of the maximum-wavelength photon.

SET UP: The energy difference is equal to the energy of the photon, so $\Delta E = hc/\lambda$.

EXECUTE: **(a)** Using the photon wavelength to find the energy difference gives
$$\Delta E = hc/\lambda = (4.136\times10^{-15}\ \text{eV}\cdot\text{s})(3.00\times10^8\ \text{m/s})/(1.11\times10^{-6}\ \text{m}) = 1.12\ \text{eV}.$$

(b) A wavelength of $1.11\ \mu\text{m} = 1110\ \text{nm}$ is in the infrared, shorter than that of visible light.

EVALUATE: Since visible photons have more than enough energy to excite electrons from the valence to the conduction band, visible light will be absorbed, which makes silicon opaque.

42.19. **IDENTIFY:** This problem deals with the density of states $g(E)$.

SET UP: The original energy is 1.60 eV and we want to know the new energy. From Eq. (42.15) we see that $g(E)$ is proportional to the square root of E.

EXECUTE: **(a)** $g(E)$ is doubled. Since $g(E)$ is proportional to the square root of the energy, E must increase by a factor of 4 to double $g(E)$, so $E = 4(1.60\ \text{eV}) = 6.40\ \text{eV}$.

(b) $g(E)$ is halved. We need to decrease $g(E)$ by a factor of $\frac{1}{2}$, so E must decrease by a factor of $\frac{1}{4}$, so $E = (1.60\ \text{eV})/4 = 0.400\ \text{eV}$.

EVALUATE: A larger fractional change in E is required than the fractional change in $g(E)$.

42.21. **IDENTIFY** and **SET UP:** The electron contribution to the molar heat capacity at constant volume of a metal is $C_V = \left(\dfrac{\pi^2 kT}{2E_F}\right)R$.

EXECUTE: **(a)** $C_V = \dfrac{\pi^2 (1.381\times10^{-23}\ \text{J/K})(300\ \text{K})}{2(5.48\ \text{eV})(1.602\times10^{-19}\ \text{J/eV})}R = 0.0233R.$

(b) The electron contribution found in part (a) is $0.0233R = 0.194\ \text{J/mol}\cdot\text{K}$. This is $0.194/25.3 = 7.67\times10^{-3} = 0.767\%$ of the total C_V.

EVALUATE: **(c)** Only a small fraction of C_V is due to the electrons. Most of C_V is due to the vibrational motion of the ions.

42.23. **IDENTIFY:** The probability is given by the Fermi-Dirac distribution.

SET UP: The Fermi-Dirac distribution is $f(E) = \dfrac{1}{e^{(E-E_F)/kT} + 1}$.

EXECUTE: We calculate the value of $f(E)$, where $E = 8.520$ eV, $E_F = 8.500$ eV,

$k = 1.38 \times 10^{-23}$ J/K $= 8.625 \times 10^{-5}$ eV/K, and $T = 20°C = 293$ K. The result is $f(E) = 0.312 = 31.2\%$.

EVALUATE: Since the energy is close to the Fermi energy, the probability is quite high that the state is occupied by an electron.

42.25. **IDENTIFY:** Use the Fermi-Dirac distribution $f(E) = \dfrac{1}{e^{(E - E_F)/kT} + 1}$. Solve for $E - E_F$.

SET UP: $e^{(E - E_F)/kT} = \dfrac{1}{f(E)} - 1$.

The problem states that $f(E) = 4.4 \times 10^{-4}$ for E at the bottom of the conduction band.

EXECUTE: $e^{(E - E_F)/kT} = \dfrac{1}{4.4 \times 10^{-4}} - 1 = 2.272 \times 10^3$.

$E - E_F = kT \ln(2.272 \times 10^3) = (1.3807 \times 10^{-23}$ J/T$)(300$ K$)\ln(2.272 \times 10^3) = 3.201 \times 10^{-20}$ J $= 0.20$ eV.

$E_F = E - 0.20$ eV; the Fermi level is 0.20 eV below the bottom of the conduction band.

EVALUATE: The energy gap between the Fermi level and bottom of the conduction band is large compared to kT at $T = 300$ K and as a result $f(E)$ is small.

42.27. **IDENTIFY:** Knowing the saturation current of a *p-n* junction at a given temperature, we want to find the current at that temperature for various voltages.

SET UP: $I = I_S(e^{eV/kT} - 1)$.

EXECUTE: **(a)** (i) For $V = 1.00$ mV, $\dfrac{eV}{kT} = \dfrac{(1.602 \times 10^{-19} \text{ C})(1.00 \times 10^{-3} \text{ V})}{(1.381 \times 10^{-23} \text{ J/K})(290 \text{ K})} = 0.0400$.

$I = (0.500$ mA$)(e^{0.0400} - 1) = 0.0204$ mA.

(ii) For $V = -1.00$ mV, $\dfrac{eV}{kT} = -0.0400$. $I = (0.500$ mA$)(e^{-0.0400} - 1) = -0.0196$ mA.

(iii) For $V = 100$ mV, $\dfrac{eV}{kT} = 4.00$. $I = (0.500$ mA$)(e^{4.00} - 1) = 26.8$ mA.

(iv) For $V = -100$ mV, $\dfrac{eV}{kT} = -4.00$. $I = (0.500$ mA$)(e^{-4.00} - 1) = -0.491$ mA.

EVALUATE: **(b)** For small V, between ± 1.00 mV, $R = V/I$ is approximately constant and the diode obeys Ohm's law to a good approximation. For larger V the deviation from Ohm's law is substantial.

42.29. **IDENTIFY and SET UP:** The voltage-current relation is given by $I = I_s(e^{eV/kT} - 1)$. Use the current for $V = +15.0$ mV to solve for the constant I_s.

EXECUTE: **(a)** Find I_s: $V = +15.0 \times 10^{-3}$ V gives $I = 9.25 \times 10^{-3}$ A.

$\dfrac{eV}{kT} = \dfrac{(1.602 \times 10^{-19} \text{ C})(15.0 \times 10^{-3} \text{ V})}{(1.381 \times 10^{-23} \text{ J/K})(300 \text{ K})} = 0.5800$.

$I_s = \dfrac{I}{e^{eV/kT} - 1} = \dfrac{9.25 \times 10^{-3} \text{ A}}{e^{0.5800} - 1} = 1.177 \times 10^{-2} = 11.77$ mA.

Then can calculate I for $V = 10.0$ mV: $\dfrac{eV}{kT} = \dfrac{(1.602 \times 10^{-19} \text{ C})(10.0 \times 10^{-3} \text{ V})}{(1.381 \times 10^{-23} \text{ J/K})(300 \text{ K})} = 0.3867.$

$I = I_s(e^{eV/kT} - 1) = (11.77 \text{ mA})(e^{0.3867} - 1) = 5.56 \text{ mA.}$

(b) $\dfrac{eV}{kT}$ has the same magnitude as in part (a) but now V is negative so $\dfrac{eV}{kT}$ is negative.

$V = -15.0 \text{ mV} : \dfrac{eV}{kT} = -0.5800$ and $I = I_s(e^{eV/kT} - 1) = (11.77 \text{ mA})(e^{-0.5800} - 1) = -5.18 \text{ mA.}$

$V = -10.0 \text{ mV} : \dfrac{eV}{kT} = -0.3867$ and $I = I_s(e^{eV/kT} - 1) = (11.77 \text{ mA})(e^{-0.3867} - 1) = -3.77 \text{ mA.}$

EVALUATE: There is a directional asymmetry in the current, with a forward-bias voltage producing more current than a reverse-bias voltage of the same magnitude, but the voltage is small enough for the asymmetry not be pronounced.

42.33. **IDENTIFY** and **SET UP:** From Chapter 21, the electric dipole moment is $p = qd$, where the dipole consists of charges $\pm q$ separated by distance d.

EXECUTE: **(a)** Point charges $+e$ and $-e$ separated by distance d, so

$$p = ed = (1.602 \times 10^{-19} \text{ C})(0.24 \times 10^{-9} \text{ m}) = 3.8 \times 10^{-29} \text{ C} \cdot \text{m.}$$

(b) $p = qd$, so $q = \dfrac{p}{d} = \dfrac{3.0 \times 10^{-29} \text{ C} \cdot \text{m}}{0.24 \times 10^{-9} \text{ m}} = 1.3 \times 10^{-19} \text{ C.}$

(c) $\dfrac{q}{e} = \dfrac{1.3 \times 10^{-19} \text{ C}}{1.602 \times 10^{-19} \text{ C}} = 0.81.$

(d) $q = \dfrac{p}{d} = \dfrac{1.5 \times 10^{-30} \text{ C} \cdot \text{m}}{0.16 \times 10^{-9} \text{ m}} = 9.37 \times 10^{-21} \text{ C.}$

$$\dfrac{q}{e} = \dfrac{9.37 \times 10^{-21} \text{ C}}{1.602 \times 10^{-19} \text{ C}} = 0.058.$$

EVALUATE: The fractional ionic character for the bond in HI is much less than the fractional ionic character for the bond in NaCl. The bond in HI is mostly covalent and not very ionic.

42.35. **IDENTIFY:** $E_{ex} = \dfrac{L^2}{2I} = \dfrac{\hbar^2 l(l+1)}{2I}.$ $E_g = 0$ ($l = 0$), and there is an additional multiplicative factor of $2l + 1$ because for each l state there are really $(2l + 1)$ m_l states with the same energy.

SET UP: From Example 42.2, $I = 1.449 \times 10^{-46} \text{ kg} \cdot \text{m}^2$ for CO.

EXECUTE: **(a)** $\dfrac{n_l}{n_0} = (2l + 1)e^{-\hbar^2 l(l+1)/(2IkT)}.$

(b) (i) $E_{l=1} = \dfrac{\hbar^2(1)(1+1)}{2(1.449 \times 10^{-46} \text{ kg} \cdot \text{m}^2)} = 7.67 \times 10^{-23} \text{ J.}$ $\dfrac{E_{l=1}}{kT} = \dfrac{7.67 \times 10^{-23} \text{ J}}{(1.38 \times 10^{-23} \text{ J/K})(300 \text{ K})} = 0.0185.$

$(2l + 1) = 3$, so $\dfrac{n_{l=1}}{n_0} = (3)e^{-0.0185} = 2.95.$

(ii) $\dfrac{E_{l=2}}{kT} = \dfrac{\hbar^2(2)(2+1)}{2(1.449 \times 10^{-46} \text{ kg} \cdot \text{m}^2)(1.38 \times 10^{-23} \text{ J/K})(300 \text{ K})} = 0.0556.$ $(2l + 1) = 5$, so

$$\frac{n_{l=1}}{n_0} = (5)(e^{-0.0556}) = 4.73.$$

(iii) $\dfrac{E_{l=10}}{kT} = \dfrac{\hbar^2 (10)\,(10+1)}{2(1.449\times10^{-46}\ \text{kg}\cdot\text{m}^2)(1.38\times10^{-23}\ \text{J/K})(300\ \text{K})} = 1.02.$

$$(2l+1) = 21,\ \text{so}\ \frac{n_{l=10}}{n_0} = (21)(e^{-1.02}) = 7.57.$$

(iv) $\dfrac{E_{l=20}}{kT} = \dfrac{\hbar^2 (20)(20+1)}{2(1.449\times10^{-46}\ \text{kg}\cdot\text{m}^2)(1.38\times10^{-23}\ \text{J/K})(300\ \text{K})} = 3.90.\ (2l+1) = 41,\ \text{so}$

$$\frac{n_{l=20}}{n_0} = (41)e^{-3.90} = 0.833.$$

(v) $\dfrac{E_{l=50}}{kT} = \dfrac{\hbar^2 (50)(50+1)}{2(1.449\times10^{-46}\ \text{kg}\cdot\text{m}^2)(1.38\times10^{-23}\ \text{J/K})(300\ \text{K})} = 23.7.\ (2l+1) = 101,\ \text{so}$

$$\frac{n_{l=50}}{n_0} = (101)e^{-23.7} = 5.38\times10^{-9}.$$

EVALUATE: **(c)** There is a competing effect between the $(2l+1)$ term and the decaying exponential. The $2l+1$ term dominates for small l, while the exponential term dominates for large l.

42.37. **IDENTIFY:** This problem is about the power in a diode.

SET UP: We use Eq. (42.22) and $P = IV$.

EXECUTE: **(a)** We want the power.
$$P = IV = I_S\left(e^{eV/kT} - 1\right)V = (1.2\times10^{-11}\ \text{A})\left(e^{e(0.6\ \text{V})/k(300\ \text{K})} - 1\right)(0.6V) = 85\ \text{mW}.$$

(b) We want V when $P = 500$ mA $= 0.5$ A. We can solve this by trial-and-error. First put the equation in the convenient form
$$P = (1.2\times10^{-11}\ \text{A})\left(e^{(38.65\ \text{V}^{-1})V} - 1\right)V$$

We know that V must be greater than 0.6 V. Trying $V = 0.8$ V gives $P = 260$ W. Trying $V = 0.7$ V gives $P = 4.7$ W. Trying $V = 0.65$ V gives $P = 635$ mW. Trying $V = 0.64$ V gives $P = 425$ mW. Trying $V = 0.645$ V gives $P = 520$ mW, which is close enough to 500 mA. So the maximum voltage is $V = 0.645$ V. To one significant figure, the answer is $V = 0.6$ V.

(c) We want the current when $V = 0.645$ V. $I = P/V = (520\ \text{mW})/(0.645\ \text{V}) = 0.81$ A.

EVALUATE: Note in part (b) how sensitive the power is to small changes in V due to the exponential.

42.39. **IDENTIFY:** The vibrational energy levels are given by $E_n = (n+\tfrac{1}{2})\,\hbar\sqrt{\dfrac{k'}{m_r}}$. The zero-point energy

is
$$E_0 = \tfrac{1}{2}\hbar\sqrt{\frac{2k'}{m_H}}.$$

SET UP: For H_2, $m_r = \dfrac{m_H}{2}$.

EXECUTE: $E_0 = \tfrac{1}{2}(1.054\times10^{-34}\ \text{J}\cdot\text{s})\sqrt{\dfrac{2(576\ \text{N/m})}{1.67\times10^{-27}\ \text{kg}}} = 4.38\times10^{-20}\ \text{J} = 0.274\ \text{eV}.$

EVALUATE: This is much less than the magnitude of the H_2 bond energy.

42.41. **IDENTIFY** and **SET UP:** Use $I = m_r r_0^2$ to calculate I. The energy levels are given by

$E_{nl} = l(l+1)\left(\dfrac{\hbar^2}{2I}\right) + (n + \tfrac{1}{2})\hbar\sqrt{\dfrac{k'}{m_r}}$. The transition energy ΔE is related to the photon wavelength by

$\Delta E = hc/\lambda$.

EXECUTE: **(a)** $m_r = \dfrac{m_H m_I}{m_H + m_I} = \dfrac{(1.67\times10^{-27}\text{ kg})(2.11\times10^{-25}\text{ kg})}{1.67\times10^{-27}\text{ kg} + 2.11\times10^{-25}\text{ kg}} = 1.657\times10^{-27}\text{ kg}.$

$\qquad I = m_r r_0^2 = (1.657\times10^{-27}\text{ kg})(0.160\times10^{-9}\text{ m})^2 = 4.24\times10^{-47}\text{ kg}\cdot\text{m}^2.$

(b) The energy levels are $E_{nl} = l(l+1)\left(\dfrac{\hbar^2}{2I}\right) + (n + \tfrac{1}{2})\hbar\sqrt{\dfrac{k'}{m_r}}.$

$\qquad\qquad \sqrt{\dfrac{k'}{m}} = \omega = 2\pi f$, so $E_{nl} = l(l+1)\left(\dfrac{\hbar^2}{2I}\right) + (n + \tfrac{1}{2})hf.$

(i) Transition $n = 1 \rightarrow n = 0, l = 1 \rightarrow l = 0$:

$$\Delta E = (2-0)\left(\dfrac{\hbar^2}{2I}\right) + (1 + \tfrac{1}{2} - \tfrac{1}{2})hf = \dfrac{\hbar^2}{I} + hf.$$

$$\Delta E = \dfrac{hc}{\lambda}, \text{ so } \lambda = \dfrac{hc}{\Delta E} = \dfrac{hc}{(\hbar^2/I) + hf} = \dfrac{c}{(\hbar/2\pi I) + f}.$$

$$\dfrac{\hbar}{2\pi I} = \dfrac{1.055\times10^{-34}\text{ J}\cdot\text{s}}{2\pi(4.24\times10^{-47}\text{ kg}\cdot\text{m}^2)} = 3.960\times10^{11}\text{ Hz}.$$

$$\lambda = \dfrac{c}{(\hbar/2\pi I) + f} = \dfrac{2.998\times10^8\text{ m/s}}{3.960\times10^{11}\text{ Hz} + 6.93\times10^{13}\text{ Hz}} = 4.30\ \mu\text{m}.$$

(ii) Transition $n = 1 \rightarrow n = 0, l = 2 \rightarrow l = 1$:

$$\Delta E = (6-2)\left(\dfrac{\hbar^2}{2I}\right) + hf = \dfrac{2\hbar^2}{I} + hf.$$

$$\lambda = \dfrac{c}{2(\hbar/2\pi I) + f} = \dfrac{2.998\times10^8\text{ m/s}}{2(3.960\times10^{11}\text{ Hz}) + 6.93\times10^{13}\text{ Hz}} = 4.28\ \mu\text{m}.$$

(iii) Transition $n = 2 \rightarrow n = 1, l = 2 \rightarrow l = 3$:

$$\Delta E = (6-12)\left(\dfrac{\hbar^2}{2I}\right) + hf = -\dfrac{3\hbar^2}{I} + hf.$$

$$\lambda = \dfrac{c}{-3(\hbar/2\pi I) + f} = \dfrac{2.998\times10^8\text{ m/s}}{-3(3.960\times10^{11}\text{ Hz}) + 6.93\times10^{13}\text{ Hz}} = 4.40\ \mu\text{m}.$$

EVALUATE: The vibrational energy change for the $n = 1 \rightarrow n = 0$ transition is the same as for the $n = 2 \rightarrow n = 1$ transition. The rotational energies are much smaller than the vibrational energies, so the wavelengths for all three transitions don't differ much.

42.47. **IDENTIFY** and **SET UP:** The occupation probability $f(E)$ is $f(E) = \dfrac{1}{e^{(E-E_F)/kT} + 1}$.

EXECUTE: **(a)** Figure 42.47 shows the graph of E versus $\ln\left\{[1/f(E)] - 1\right\}$ for the data given in the problem. The slope of the best-fit straight line is 0.445 eV, and the y-intercept is 1.80 eV.

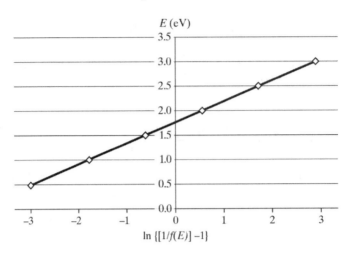

Figure 42.47

(b) Solve the $f(E)$ equation for E, giving $e^{(E-E_F)/kT} = 1/f(E) - 1$. Now take natural logarithms of both side of the equation, giving $(E - E_F)/kT = \ln\{[1/f(E)] - 1\}$, which gives

$E = kT\ln\{[1/f(E)] - 1\} + E_F$. From this we see that a graph of E versus $\ln\{[1/f(E)] - 1\}$ should be a straight line having a slope equal to kT and a y-intercept equal to E_F. From our graph, we get $E_F = y$-intercept = 1.80 eV.

The slope is equal to kT, so $T = (\text{slope})/k = (0.445 \text{ eV})/(1.38 \times 10^{-23} \text{ J/K})$
$$= (7.129 \times 10^{-20} \text{ J})/(1.38 \times 10^{-23} \text{ J/K})$$
$$= 5170 \text{ K}.$$

EVALUATE: In Example 42.7, the Fermi energy for copper was found to be 7.03 eV, so our result of 1.80 eV seems plausible.

42.49. **IDENTIFY:** This problem involves a semiconductor device used to make a half-wave rectifier.

SET UP: Eq. (42.22) gives the current through the junction of the semiconductor. The saturation current is 5×10^{-13} A and $T = 300$ K.

EXECUTE: **(a)** We want the minimum V_{out}, so $I = 1$ μA. Solve Eq. (42.22) for V. (Careful! In this equation, e is the charge of an electron, not the base for natural logarithms.)

$$V_{out} = \frac{kT}{e}\ln\left(1 + \frac{I}{I_S}\right) = \frac{k(300 \text{ K})}{e}\ln\left(1 + \frac{1 \mu A}{5 \times 10^{-13} \text{ A}}\right) = 0.375 \text{ V}.$$

(b) We want the maximum V_{out}. Combine Eq. (42.22) and $I = V/R_D$ and solve for V. Note that $e/kT = e/[k(300 \text{ K})] = 38.65$ V^{-1}.

$$V = R_D I = R_D I_S\left(e^{eV/kT} - 1\right) = (100 \text{ }\Omega)(5 \times 10^{-13} \text{ A})\left(e^{(38.65 \text{ V}^{-1})V} - 1\right).$$

Solve this equation by trial-and-error or using mathematical software. The result is $V_{max} = 0.6$ V.

(c) We want the current. $I = V/R_D = (0.6 \text{ V})/(100 \text{ }\Omega) = 0.006$ A.

(d) If $V_{in} > V_{max}$: $V_{out} = V_{max} = 0.6$ V.
If $V_{in} < V_{max}$: $V_{out} = V_{in}$.

EVALUATE: The exponential behavior makes I very sensitive to changes in V.

STUDY GUIDE FOR NUCLEAR PHYSICS

Summary

In this chapter, we look deep inside the atom into the nucleus and the particles that make up the nucleus. We'll investigate how protons and neutrons combine to influence the size, mass, and stability of a nucleus. We'll then turn to radioactive decay and nuclear reactions to examine the processes that shape our understanding of nuclei. Beneficial uses of radioactivity, fission, and fusion will also be examined.

Objectives

After studying this chapter, you'll understand

- How protons and neutrons form nuclei.
- Some fundamental properties of nuclei, including radii, densities, spins, and magnetic moments.
- How binding energies depend on the number of protons and neutrons in the nucleus.
- How nuclei undergo radioactive decay and how they decay at different rates.
- The basics of nuclear reactions and how to predict reaction energies.

Concepts and Equations

Term	Description
Properties of the Nucleus	A nucleus is composed of A nucleons (Z protons and N neutrons), is roughly spherical in shape, and has a radius that depends on A according to the formula $$R = R_0 A^{1/3},$$ where $R_0 = 1.2$ fm.
Nuclear Binding and Structure	The mass of the nucleus is less than the total mass of the protons and neutrons that constitute the nucleus. The mass difference multiplied by c^2 is the binding energy E_B, given by $$E_B = (ZM_H + Nm_n - {}^A_Z M)c^2,$$ where z is the number of protons in the nucleus, M_H is atomic mass of neutral hydrogen, N is the number of neutrons in the nucleus, m_n is the mass of a neutron, ${}^A_Z M$ is the mass of the neutral atom containing the nucleus, and c is the speed of light in vacuum. Nuclei are held together by the attractive nuclear force overcoming the repulsive electrical force. The nuclear force favors paired nucleons with opposite spin and pairs of pairs. Most stable nuclei have more neutrons than protons. Both the shell model and the liquid drop model are used to describe the properties of the nucleus.
Radioactivity and Radiation	Unstable nuclei undergo radioactive decay, most commonly through alpha (α) and beta (β) particle emission and sometimes through gamma-ray emission. Alpha particles are two protons and two neutrons bound together. The most common beta particles (beta-minus particles) are electrons. Radioactive material decays exponentially. The number N of nuclei remaining in a sample of N_0 nuclei after time t is $$N = N_0 e^{-\lambda t},$$ where λ is the decay constant for the particular nuclear species in question. The rate of decay is described by the decay constant λ, the half-life $T_{1/2}$, or the lifetime T_{mean}, related by $$T_{mean} = \frac{1}{\lambda} = \frac{T_{1/2}}{\ln 2} = \frac{T_{1/2}}{0.693}.$$ Radioactivity can be used beneficially to date artifacts or to diagnose and treat medical conditions. Radioactivity also can be harmful to human tissue, as it can ionize cellular material, but radiation hazards can be reduced through proper precautions.
Nuclear Reactions, Fission, and Fusion	Nuclear reactions result from the bombardment of a nucleus by a particle. Energy is exchanged in nuclear reactions. A reaction resulting in an excess of kinetic energy is called an exoergic reaction, and a reaction resulting in a deficiency of kinetic energy is called an endoergic reaction. Fission is the radioactive decay of an unstable nucleus into two or more nuclei. Fission is used to power nuclear reactors. Fusion is a nuclear reaction in which two or more light nuclei combine to form a larger nucleus plus excess energy.

Key Concept 1: The radius of a nucleus is approximately proportional to the 1/3 power of the number of nucleons A that the nucleus contains. The volume is approximately proportional to A.

Key Concept 2: Both the proton and the neutron have a magnetic moment $\vec{\mu}$ associated with its spin. When placed in a magnetic field $\vec{B}$, a proton or neutron has two states of different energy: one where the component of $\vec{\mu}$ in the direction of $\vec{B}$ is positive and one where it is negative.

Key Concept 3: The mass defect of a nucleus is the difference between the sum of the masses of its nucleons and the mass of the nucleus itself. We calculate this using the mass of a hydrogen atom rather than the mass of a proton, and using the mass of a neutral atom that contains the nucleus rather than the nuclear mass. To find the binding energy, multiply the mass defect by c^2; to then find the binding energy per nucleon, divide by the nucleon number A.

Key Concept 4: You can use the liquid-drop model and its associated semi-empirical mass formula to calculate the binding energy and mass of a nuclide.

Key Concept 5: In alpha decay, a large, unstable nucleus with Z protons and N neutrons emits an alpha particle (a $^{4}_{2}$He nucleus). The daughter nucleus has $Z - 2$ protons and $N - 2$ neutrons. The combined mass of the decay products is less than the mass of the parent; the lost rest energy goes into kinetic energy of the decay products.

Key Concept 6: In beta-minus decay, an unstable nucleus with Z protons and N neutrons decays by converting a neutron into a proton and emitting an electron and an antineutrino. The daughter nucleus has $Z + 1$ protons and $N - 1$ neutrons. This is possible only if the neutral parent atom has a greater mass than the neutral daughter atom.

Key Concept 7: In beta-plus decay, an unstable nucleus with Z protons and N neutrons decays by converting a proton into a neutron and emitting a positron and a neutrino. The daughter nucleus has $Z - 1$ protons and $N + 1$ neutrons. This is possible only if the neutral parent atom has a greater mass than the neutral daughter atom plus two electrons. If this is not the case, electron capture may be possible.

Key Concept 8: Because radioactive decay is a statistical process, the number of radioactive nuclei $N(t)$ present in a radioactive sample and the activity $-dN(t)/dt$ of the sample both decay exponentially with time. Both $N(t)$ and $-dN(t)/dt$ decrease by ½ in one half-life and by $1/e$ in one mean lifetime.

Key Concept 9: To find the age of a sample using radioactive dating, measure its present activity and determine what its activity would have been when it formed. The age is proportional to the logarithm of the ratio of these two activities.

Key Concept 10: The absorbed dose for a sample of tissue exposed to radiation is the amount of energy delivered per kilogram to the tissue. The equivalent dose equals the absorbed dose multiplied by the relative biological effectiveness (RBE), which depends on the type of radiation used.

Key Concept 11: To find the energy released in a nuclear reaction (the reaction energy), calculate the difference between the total mass of the initial nuclei and the total mass of the final nuclei and then multiply the result by c^2.

Key Concept 12: The thermal power output of a power plant equals the energy released per reaction multiplied by the number of reactions that take place per second. Since the energy released per nuclear fission reaction is very large, a substantial power output can be obtained from relatively few fission reactions per second.

Key Concept 13: When two light nuclei undergo fusion to form a more massive nucleus, the binding energy per nucleon increases. Hence energy is released in such fusion reactions.

Conceptual Questions

1: Size of nuclei

How much must the mass number of a nucleus increase to double the volume of the nucleus? To double its radius?

From Chapter 43 of Student's Study Guide to accompany *University Physics with Modern Physics, Volume 3*, Fifteenth Edition. Hugh D. Young and Roger A. Freedman. Copyright © 2020 by Pearson Education, Inc. All rights reserved.

IDENTIFY, SET UP, AND EXECUTE The radius of a nucleus is proportional to the cube root of the mass number. Volume is proportional to the cube of the radius. To double the volume of a nucleus, the mass number must double.

To double the radius, the mass number must increase by a factor of $2^3 = 8$.

EVALUATE Changes in radius and volume give indications of the structure of nuclei. The relation of the radius to the volume indicates that the nuclear density is roughly constant for all nuclei.

2: Causing nuclear fusion

Deuterium nuclei can fuse, liberating energy. Why doesn't a container of deuterium gas begin fusion after being shaken?

IDENTIFY, SET UP, AND EXECUTE The deuterium nuclei can fuse, but only when the nuclei contact each other. Deuterium nuclei in deuterium gas are surrounded by electron clouds, and Coulomb repulsion prevents the nuclei from coming into contact with each other. For the fusion reaction to take place, the deuterium nuclei must be given enough energy to overcome the Coulomb repulsion.

EVALUATE The biggest barrier to fusion reactions being used as a source of energy is the electric repulsion between charged nuclei. Even bare nuclei without electron clouds must have enough kinetic energy to overcome that repulsion in order for fusion to occur.

Problems

1: Binding energy of silver and gold

(a) Calculate the total binding energy of ^{107}Ag (atomic mass 106.905097 u) and ^{197}Au (atomic mass 196.966569 u). (b) Calculate the binding energy per nucleon for each atom. (c) Which of these nuclei is more tightly bound?

IDENTIFY We find the binding energy by converting the mass defect into energy.

SET UP The nucleon number is the number of nucleons, so we'll divide the binding energy by 107 and 197 to find the respective binding energies per nucleon.

EXECUTE The mass defect is calculated by subtracting the atomic mass from the combined mass of the protons and neutrons in the nucleus:

$$\Delta M = Zm_p + Nm_n - M = Z(1.007825 \text{ u}) + N(1.008665 \text{ u}) - M.$$

^{107}Ag has 47 protons and 60 neutrons, for a total of 107 nucleons. Its mass defect is

$$\Delta M = (47)(1.007825 \text{ u}) + (60)(1.008665 \text{ u}) - (106.905097 \text{ u}) = 0.9826 \text{ u}.$$

Its binding energy is

$$E_B = (0.9826 \text{ u})(931.5 \text{ MeV/u}) = 915.3 \text{ MeV}.$$

Its binding energy per nucleon is

$$E_B/A = 915.3 \text{ MeV}/107 = 8.55 \text{ MeV/nucleon}.$$

^{197}Au has 79 protons and 118 neutrons, for a total of 197 nucleons. Its mass defect is

$$\Delta M = (79(1.007825 \text{ u}) + (118)(1.008665 \text{ u}) - (196.966569 \text{ u}) = 1.6741 \text{ u}.$$

Its binding energy is

$$E_B = (1.6741 \text{ u})(931.5 \text{ MeV/u}) = 1559.4 \text{ MeV}$$

Its binding energy per nucleon is

$$E_B/A = 1559.4 \text{ MeV}/197 = 7.92 \text{ MeV/nucleon}.$$

The silver atom's nucleons are more tightly bound, since they require more energy per nucleon to become free.

KEY CONCEPT 3 **EVALUATE** The total binding energy is larger for gold than for silver, but the binding energy per nucleon is less for gold. This state of affairs agrees with our model of the nucleus: Atoms with larger numbers of nucleons have greater radii, causing their nucleons to be less strongly bound than atoms with smaller numbers of nucleons.

Practice Problem: What is the binding energy of radium-226 (atomic mass 226.025410 u)? *Answer:* 1731.6 MeV

Extra Practice: What is the binding energy per nucleon of radium-226? *Answer:* 7.66 MeV/nucleon

2: Examining a nuclear reaction

Calculate the energy released or absorbed by the reaction $^3_2\text{He} + ^2_1\text{H} \rightarrow ^4_2\text{He} + ^1_1\text{H}$.

IDENTIFY AND SET UP We find the reaction energy by subtracting the rest masses of the products from the rest masses of the initial particles. Table 2 in the text provides the needed rest masses.

EXECUTE The reaction energy is

$$Q = (M_A + M_B - M_C - M_D)c^2,$$

where M_A and M_B are the rest masses of ^3_2He and ^2_1H, respectively, and M_C and M_D are the rest masses of ^4_2He and ^1_1H, respectively. Table 2 in the text gives the rest masses:

^3_2He: 3.016029 u,

^2_1H: 2.014101 u,

^4_2He: 4.002603 u,

^1_1H: 1.007825 u.

The change in mass is

$$Q = (3.016029 \text{ u}) + (2.014101 \text{ u}) - (4.002603 \text{ u}) - (1.007825 \text{ u}) = 0.0197 \text{ u}.$$

The mass has decreased, indicating that energy is released. The reaction energy is

$$Q = (0.0197 \text{ u})(931.5 \text{ MeV/u}) = 18.35 \text{ MeV}.$$

A total of 18.35 MeV is released in the reaction.

KEY CONCEPT 11 **EVALUATE** Since the total mass decreases, the kinetic energy of the products is more than the initial energy of the interacting nuclei. This fusion reaction is one candidate being considered for the production of fusion energy.

We also should check that charge is conserved in the reaction. The charge of the incoming nuclei is $+3e$, and the charge of the product nuclei is $+3e$, confirming that charge is conserved in the reaction.

Practice Problem: What is the reaction energy of two deuterons combining to form helium-4? *Answer:* 23.8 MeV

Extra Practice: What is the reaction energy of the fictional case of carbon-12 splitting into 3 helium-4 atoms? *Answer:* −7.27 MeV

3: Radioactivity

At a certain time, a sample of radioactive material is measured and is found to decay at a rate of 32 counts per second. Two hours later, the sample is measured to decay at a rate of 13 counts per second. What is the half-life of the sample?

IDENTIFY AND SET UP Radioactivity is an exponential decay process. The two decay rates have the same decay constant. We will divide the initial rate by the later rate and substitute the resulting ratio into the radioactive decay formula to find the decay constant. The half-life is then found from the decay constant.

EXECUTE The initial decay rate can be written

$$\frac{\Delta N}{\Delta t} = 32 \text{ counts/s} = -\lambda N_0.$$

The later decay rate can be written

$$\frac{\Delta N}{\Delta t} = 13 \text{ counts/s} = -\lambda N_2.$$

We divide the second decay rate by the first decay rate to find the ratio N_2/N_0:

$$\frac{N_2}{N_0} = \frac{13 \text{ counts/s}}{32 \text{ counts/s}} = 0.4063.$$

The number of counts at 2 hours must be the initial number of counts times $e^{-e\lambda t}$:

$$N_2 = N_0 e^{-\lambda t} = N_0 e^{-\lambda(2 \text{ hours})}.$$

We substitute and rearrange terms to solve for λ:

$$\frac{N_2}{N_0} = 0.4063 = e^{-\lambda(7200 \text{ s})},$$

$$\ln 0.4063 = \ln(e^{-\lambda(7200 \text{ s})}) = -\lambda(7200 \text{ s}),$$

$$\lambda = -\frac{\ln 0.4063}{7200 \text{ s}} = 1.25 \times 10^{-4}/\text{s}.$$

The half-life is then

$$T_{1/2} = \frac{\ln 2}{\lambda} = \frac{\ln 2}{1.25 \times 10^{-4}/\text{s}} = 5540 \text{ s} = 1.54 \text{ hr}.$$

The half-life of the sample is 1.54 hours.

KEY CONCEPT 8 **EVALUATE** We check our result by noting that the decay rate dropped by more than a factor of 2 in 1 hour, indicating that the half-life was less than 2 hours, in agreement with our result.

From Chapter 43 of Student's Study Guide to accompany *University Physics with Modern Physics, Volume 3,* Fifteenth Edition. Hugh D. Young and Roger A. Freedman. Copyright © 2020 by Pearson Education, Inc. All rights reserved.

Practice Problem: How long will it take for the sample to reach 4 counts/second? *Answer:* 4.62 hours

Extra Practice: How long will it take for 99% of the sample to decay? *Answer:* 10.2 hours

4: Carbon dating

The half-life of ^{14}C is 5568 years. If ^{14}C dating was done on a piece of 2000-year-old wood, how would the abundance of ^{14}C in the wood compare with that of ^{14}C in freshly cut wood from a similar tree?

IDENTIFY AND SET UP ^{14}C is a radioactive nucleus, so its mass decreases exponentially with time. We'll use the half-life formula to solve the problem.

EXECUTE The number of nuclei decay according to

$$N = N_0 e^{-\lambda t}.$$

The decay constant for ^{14}C is

$$\lambda = \frac{0.693}{T_{1/2}} = 1.24 \times 10^{-4}/\text{yr}.$$

We'll take N_0 as the number of ^{14}C nuclei in the new wood. N is the number of ^{14}C nuclei in the old wood. Their ratio is

$$\frac{N}{N_0} = e^{-\lambda t} = e^{-(1.24 \times 10^{-4}/\text{yr})(2000\text{ yr})} = 0.78.$$

The old wood has 78% of the abundance of ^{14}C in the new wood.

KEY CONCEPT 9 **EVALUATE** If, instead, we were interested in the abundance of ^{14}C in the old wood, we could have found the age of that wood.

Try It Yourself!

1: ^{14}N mass and energy

Find the mass defect, the total binding energy, and the binding energy per nucleon for ^{14}N.

Solution Checkpoints

IDENTIFY AND SET UP: Use tables in the text and mass relations to solve.

EXECUTE: The mass of the ^{14}N nucleus is

$$m(^{14}\text{N}) = 14.00307\text{ u} - 7(0.000549\text{ u}) = 13.99923\text{ u}.$$

The combined mass of seven protons and seven neutrons is 14.11159 u, so the mass defect is 0.124 u.

The mass defect is the binding energy divided by c^2, so the binding energy is

$$E_B = (0.124\text{ u})(931.5\text{ MeV/u}) = 104.7\text{ MeV}.$$

Dividing the binding energy by the 14 nucleons gives a binding energy per nucleon of 7.5 MeV.

From Chapter 43 of Student's Study Guide to accompany *University Physics with Modern Physics, Volume 3,* Fifteenth Edition. Hugh D. Young and Roger A. Freedman. Copyright © 2020 by Pearson Education, Inc. All rights reserved.

KEY CONCEPT 3 **EVALUATE** If seven protons and seven neutrons were brought together to form an ^{14}N nucleus, would energy be released or used up in the process?

> **Practice Problem:** What is the binding energy of oxygen-16 (atomic mass 15.994915 u)?
> *Answer:* 127.6 MeV

> **Extra Practice:** What is the binding energy of oxygen-17 (atomic mass 16.999132 u)?
> *Answer:* 131.8 MeV

2: $A=3$ nuclear decay

(a) Find the mass of the tritium nucleus $^3_1$H (b) Find the mass of the helium nucleus $^3_2$He plus an electron at rest. (c) If a tritium nucleus decays into a helium nucleus, plus an electron, plus a neutrino, how much energy is released as kinetic energy?

Solution Checkpoints

IDENTIFY AND SET UP Use Table 3 in the text and mass relations to solve.

EXECUTE (a) The mass of the tritium nucleus is

$$m(^3_1\text{H}) = 3.016049 \text{ u} - 1(0.000549 \text{ u}) = 3.015500 \text{ u}.$$

(b) The mass of the helium nucleus plus an electron at rest is

$$m(^3_2\text{He}) + m_e = 3.016029 \text{ u} - 2(0.000549 \text{ u}) + 0.000549 \text{ u} = 3.015480 \text{ u}.$$

(c) In the decay, the kinetic energy is shared between the electron and the neutrino. The mass defect is 0.00002 u, or 18.6 keV.

KEY CONCEPT 11 **EVALUATE** Tritium is a radioactive hydrogen isotope that emits beta rays. What is the maximum energy of the beta rays emitted?

Key Example Variation Problems

Solutions to these problems are in Chapter 43 of the Student's Solutions Manual.

Be sure to review EXAMPLES 43.3 and 43.4 (Section 43.2) before attempting these problems.

VP43.4.1 Find (a) the mass defect, (b) the total binding energy, and (c) the binding energy per nucleon of $^{16}_8$O, which has a neutral atomic mass of 15.994915 u.

VP43.4.2 The mass defect of $^{63}_{29}$Cu is 0.5919378 u. Find (a) its binding energy, (b) its binding energy per nucleon, and (c) its neutral atomic mass.

VP43.4.3 Calculate the binding energy E_B/A for the nuclides (a) $^{75}_{33}$As (mass defect 0.7005604 u), (b) $^{150}_{62}$Sm (mass defect 1.330388 u), and (c) $^{225}_{88}$Ra (mass defect 1.852094 u). (d) Figure 43.2 shows that for nuclei with nucleon numbers $A > 62$, the binding energy per nucleon E_B/A decreases as A increases. Do your results agree with this?

VP43.4.4 Use the semiempirical mass formula to calculate the total estimated binding energy E_B and the binding energy per nucleon E_B/A for the nuclides (a) ruthenium-100 ($^{100}_{44}$Ru) and (b) mercury-200 ($^{200}_{80}$Hg). (c) Are your results consistent with Fig. 43.2?

Be sure to review EXAMPLES 43.5, 43.6, and 43.7 (Section 43.3) before attempting these problems.

VP43.7.1 Household smoke detectors contain a tiny amount of radioactive americium that decays by α emission to neptunium:

$^{241}_{95}$Am $\rightarrow$ $^{237}_{93}$Np $+$ $^4_2$He. The atomic masses are 241.056827 u for $^{241}_{95}$Am, 237.048172 u for $^{237}_{93}$Np, and 4.002603 for $^4_2$He. If the $^{241}_{95}$Am nucleus is originally at rest, find (a) the total energy released in the decay, (b) the kinetic energy of the emitted a particle, and (c) the speed of the emitted a particle.

VP43.7.2 The beryllium isotope $^8_4$Be is very unstable, and rapidly decays into two α particles: $^8_4$Be $\rightarrow$ $^4_2$He $+$ $^4_2$He. The atomic masses are 8.0053051 u for $^8_4$Be and 4.0026033 u for $^4_2$He. If the $^8_4$Be nucleus is originally at rest, find (a) the kinetic energy and (b) the speed of each emitted α particle.

VP43.7.3 State whether each of the following proposed β^- decays is possible or impossible. If it is possible, find the total energy released in the decay. The atomic mass is given for each nuclide. (a) $^{12}_4$Be (12.026922 u) to $^{12}_5$B (12.014353 u); (b) $^{33}_{17}$Cl (32.977452 u) to $^{33}_{18}$Ar (32.989926 u); (c) $^{82}_{35}$Br (81.9168018 u) to $^{82}_{36}$Kr (81.9134812 u).

VP43.7.4 State whether each of the following proposed β^+ decays is possible or impossible. If it is possible, find the total energy released in the decay; if it is impossible, state whether the decay can occur by electron capture. The atomic mass is given for each nuclide. The combined mass of two electrons is 0.001097 u. (a) $^{46}_{23}$V (45.960198 u) to $^{46}_{22}$Ti (45.952627 u); (b) $^{134}_{57}$La (133.908514 u) to $^{134}_{56}$Ba (133.904508 u); (c) $^{67}_{31}$Ga (66.9282024u) to $^{67}_{30}$Zn (66.9271275 u).

Be sure to review EXAMPLES 43.8 and 43.9 (Section 43.4) before attempting these problems.

VP43.9.1 The sodium isotope $^{24}_{11}$Na undergoes β^- decay to $^{24}_{12}$Mg (a stable isotope) with a half-life of 15.0 h. Initially a sample of $^{24}_{11}$Na has β^- activity of 1.50 μCi. Find (a) the mean lifetime, (b) the decay constant, (c) the initial number of $^{24}_{11}$Na nuclei in the sample, and (d) the activity of the sample after 24.0 h.

VP43.9.2 You are given a sample that was initially 100% $^{95}_{41}$Nb. This isotope undergoes β^- decay to $^{95}_{42}$Mo (a stable isotope) with a half-life of 35.0 h. You find that the sample now contains 23.8% $^{95}_{41}$Nb and 76.2% $^{95}_{42}$Mo and has a β^- activity of 0.514 μCi. Find (a) the decay constant, (b) the present number of $^{95}_{41}$Nb nuclei in the sample, (c) the initial number of $^{95}_{41}$Nb nuclei in the sample, and (d) how much time has elapsed since the sample was 100% $^{95}_{41}$Nb.

VP43.9.3 A certain isotope undergoes γ decay from an excited state to its ground state. Initially a sample of this isotope has an activity of 0.425 μCi; 45.0 s later, the activity has decreased to 0.121 μCi. Find (a) the decay constant and half-life of the excited state and (b) the number of nuclei in the excited state initially and 45.0 s later.

VP43.9.4 Before 1900 the activity per unit mass of atmospheric carbon due to the presence of ^{14}C (half-life 5730 y) averaged about 0.255 Bq per gram of carbon. You detect 113 decays of ^{14}C in 20.0 min from a sample of carbon of mass 0.510 g. Assuming that its activity per unit mass of carbon when it formed was the pre-1900 average value for the air, find (a) the age of the sample and (b) the current number of ^{14}C nuclei in the sample.

VP43.4.1. **IDENTIFY:** This problem is about the binding energy of an O-16 nucleus.
SET UP and **EXECUTE:** **(a)** We want the mass defect ΔM. Use the numerical values given in the text and Table 43.2. For O-16, $Z = 8$ and $A = 16$ so $N = 8$.

$$\Delta M = ZM_{\mathrm{H}} + Nm_{\mathrm{n}} - {}^A_Z M.$$

$\Delta M = 8(1.007825 \text{ u}) - 8(1.008655 \text{ u}) - 15.994915 \text{ u} = 0.137005 \text{ u}.$
(b) We want the binding energy. $E_{\mathrm{B}} = \Delta Mc^2 = (0.137005 \text{ u})(931.5 \text{ MeV/u}) = 127.6 \text{ MeV}.$
(c) We want the binding energy per nucleon. O-16 has 16 nucleons, so
$$E_{\mathrm{B}}/\text{nucleon} = (127.6 \text{ MeV})/16 = 7.976 \text{ MeV/nucleon}.$$
EVALUATE: From the graph in Figure 43.2, we see that O-16 is more tightly bound that C-12 but less so than Ni-62.

VP43.4.2. **IDENTIFY:** This problem is about the energy of the Cu-63 nucleus.
SET UP and **EXECUTE:** We know that the mass defect is 0.5919378 u. **(a)** The binding energy E_{B} is the energy of the mass defect. $E_{\mathrm{B}} = (0.5919378 \text{ u})(931.5 \text{ MeV/u}) = 551.4 \text{ MeV}.$
(b) $E_{\mathrm{B}}/\text{nucleon} = (551.4 \text{ MeV})/(63 \text{ nucleons}) = 8.752 \text{ MeV/nucleon}.$
(c) We want the mass of the Cu-63 atom. Use Eq. (43.10) with $Z = 29$ and $A = 63$, so $N = 34$.
EVALUATE: Note that the C-63 mass is about 63 u, but *not quite*.

VP43.4.3. **IDENTIFY:** This problem is about nuclear binding energy.
SET UP: The binding energy per nucleon is E_{B}/A where E_{B} is the energy of the mass defect. Use the given mass defects to calculate E_{B}.
EXECUTE: **(a)** As-75. $E_{\mathrm{B}}/A = (0.7005604 \text{ u})(931.5 \text{ MeV/u})/75 = 8.701 \text{ MeV/nucleon}.$
(b) Sm-150. Use the same method as in (a) with $A = 150$, giving $E_{\mathrm{B}}/A = 8.262 \text{ MeV/nucleon}.$
(c) Ra-225. $A = 225$ gives $E_{\mathrm{B}}/A = 7.668 \text{ MeV/nucleon}.$
(d) As A increases from 75 to 225, E_{B}/A decreases from 8.701 MeV/nucleon to 7.668 MeV/nucleon. This behavior agrees with Figure 43.2.
EVALUATE: Note that the *total* binding energy increases as A increases, but the binding energy *per nucleon* decreases.

VP43.4.4. **IDENTIFY:** In this problem, we use the semiempirical mass formula to calculate the binding energy.
SET UP: Eq. (43.11) gives the formula and the constants $C_1, C_2, \ldots, C_5$ are given following it. We follow the procedure of Example 43.4.
EXECUTE: **(a)** Ru-100. $A = 100$, $Z = 44$. Using Eq. (43.11) gives the following terms.
$C_1 A = (15.75 \text{ MeV})(100) = 1575$
$-C_2 A^{2/3} = -(17.80 \text{ MeV})(100^{2/3}) = -383.5$
$-C_3 Z(Z - 1)/A^{1/3} = -(0.7100 \text{ MeV})(44)(43)/(100^{1/3}) = -289.4 \text{ MeV}$
$-C_4(A - 2Z)^2/A = -(23.69 \text{ MeV})(100 - 88)^2/(100) = -34.11 \text{ MeV}$
$+C_5 A^{-4/3} = (39 \text{ MeV})(100^{-4/3}) = 0.084 \text{ MeV}.$

Adding all these terms gives $E_B = 868.1$ MeV. $E_B/A = (868.1$ MeV$)/100 = 8.681$ MeV/nucleon.
(b) Hg-200. $A = 200$, $Z = 80$. Use the same procedure as in part (a), giving $E_B = 1584$ MeV,
$E_B/A = (1584$ MeV$)/(200) = 7.922$ MeV/nucleon.
EVALUATE: **(c)** From Figure 43.2: For $A = 100$, $E_B/A \approx 8.66$ MeV/nucleon, and we got
8.681 MeV/nucleon.
For $A = 200$, $E_B/A \approx 7.85$ MeV/nucleon, and we got 79.22 MeV/nucleon. Our results agree closely
with those in the figure.

VP43.7.1. **IDENTIFY:** This problem is about energy in radioactive decay.
SET UP and EXECUTE: **(a)** We want the energy released. Refer to the decay shown in the problem.
Using the given masses, the mass difference is 241.056827 u – (237.048172 u + 4.002603 u)
= 0.0060520 u. The energy released is the energy of this mass, which is given by (0.006052 u)
(931.5 MeV/u) = 5.637 MeV.
(b) We want the kinetic energy of the alpha particle. The rest energy of He is (4 u)(931.5 MeV/u)
= 3726 MeV, which is much greater than the energy released. Therefore we do not need to use relativity. Momentum conservation gives $m_\alpha v_\alpha = m_{Np} v_{Np}$, so $v_\alpha/v_{Np} = m_{Np}/m_\alpha$. Using the masses gives v_{Np}
= (4.003/237.0)v_α = 0.016890v_α. Energy conservation gives $K_\alpha + K_{Np} = 5.637$ MeV. Using $K = \frac{1}{2}$
mv^2 and taking the ratio of kinetic energies gives $K_\alpha/K_{Np} = 59.22$. Adding the kinetic energies gives
$K_\alpha + K_\alpha/59.22 = 5.637$ MeV. $K_\alpha = 5.543$ MeV.
(c) We want the speed of the alpha particle. Use the result from (b) and $K = \frac{1}{2} mv^2$ and solve for v,
giving $v_\alpha = 1.63 \times 10^7$ m/s.
EVALUATE: $v/c = 0.16/3.0 = 0.053$, so v is about 5% the speed of light, which means that it was
OK to neglect relativity.

VP43.7.2. **IDENTIFY:** This problem is about the nuclear decay of Be-8 to two alpha particles.
SET UP: First find the mass defect and use it to find the energy released. The alpha particles each
get one-half of this energy.
EXECUTE: **(a)** We want the kinetic energy. The mass defect is $m_{Be} - 2m_{He}$, which gives
[8.0053051 u – 2(4.0026033 u)](931.5 MeV/u) = 0.09175 MeV. Each alpha particle gets half of this,
so $K_\alpha = 0.0459$ MeV.
(b) We want the speed of the alpha particles. The rest energy of the alpha is (4.0026 u)
(931.5 MeV/u) = 3730 MeV, which is much greater than its kinetic energy. So we do not need to
use relativity. Use $K = \frac{1}{2} mv^2$ with $K = 0.0459$ MeV, giving $v_\alpha = 1.49 \times 10^6$ m/s.
EVALUATE: $v/c = 0.0149/3.0 = 0.00497$, so $v \ll c$. Our neglect of relativity was justified.

VP43.7.3. **IDENTIFY:** This problem is about β^- decay.
SET UP: β^- decay is possible if the mass of the original neutral atom is greater than the mass of the
final atom. The energy released is the energy of the difference in mass.
EXECUTE: **(a)** The mass of Be-12 is greater than the mass of B-12, so this decay is possible. The
energy released is $E = (12.026922$ u – 12.014353 u$)(931.5$ MeV/u$) = 11.71$ MeV.
(b) The mass of Ar-33 is greater than the mass of Cl-33, so this decay is not possible.
(c) The mass of Br-82 is greater than the mass of Kr-82, so this decay is possible. As in part (a),
$E = (81.9168018$ u – 81.9134812 u$)(931.5$ MeV/u$) = 3.093$ MeV.
EVALUATE: The greater the mass difference, the greater the energy that is released.

VP43.7.4. **IDENTIFY:** This problem is about β^+ decay.
SET UP: β^+ decay is possible if the mass of the original neutral atom is greater than the mass of the
final atom by twice the electron mass. The mass of the electron is 0.000548580 u, so the mass of the
original atom must be greater than that of the final atom by 2(0.000548580 u) = 0.001097160 u. The
energy released is the energy of the difference in mass minus two electron masses.
EXECUTE: **(a)** The mass difference is 45.960198 – 45.952627 u = 0.0075710 u. This difference is
greater than twice the electron mass, so the decay is possible. The energy released is
$$E = (0.0075710 \text{ u} - 0.001097160 \text{ u})(931.5 \text{ MeV/u}) = 6.030 \text{ MeV}.$$

(b) $\Delta m = 133.908514$ u $- 133.904508$ u $= 0.0040060$ u $> 2m_e$, so the decay is possible. The energy is $E = (0.0040060$ u $- 0.00109716$ u$)(931.5$ MeV/u$) = 2.710$ MeV.

(c) $\Delta m = 66.928202$ u $- 66.9271275$ u $= 0.0010745$ u $< 2m_e$, so β^+ decay is not possible. The original atom's mass is greater than that of the final atom, so electron capture is possible.

EVALUATE: If the mass of the original atom is greater than the mass of the final atom, then β^+ decay and electron capture are *both* possible, so both may occur. A number of radionuclides decay by more than one mode.

VP43.9.1. IDENTIFY: This problem deals with radioactive decay. The half-life is $T_{1/2} = 15.0$ h and the initial activity is $dN/dt = -1.60$ μCi.

SET UP and EXECUTE: **(a)** $T_{mean} = T_{1/2}/(\ln 2) = (15.0$ h$)/(\ln 2) = 21.6$ h $= 7.79 \times 10^4$ s.

(b) $\lambda = (\ln 2)/T_{1/2} = (\ln 2)/(15.0$ h$) = 4.621 \times 10^{-2}$ h$^{-1} = 1.28 \times 10^{-5}$ s^{-1}.

(c) We want the initial number N_0 of Na-24 nuclei. We know the initial activity, so we make use of that.

$$N = N_0 e^{-\lambda t}$$

$$dN/dt = -\lambda N_0 e^{-\lambda t} = -\lambda N_0$$

$$N_0 = -\frac{1}{\lambda}\frac{dN}{dt} = -\frac{1}{1.28 \times 10^{-5} \text{ s}^{-1}}(-1.50 \text{ } \mu\text{Ci})(3.70 \times 10^{10} \text{ s}^{-1}) = 4.32 \times 10^9.$$

(d) We want dN/dt after 24.0 h. From our work in part (c), we can see that

$$dN/dt = dN/dt)_0 \text{ } e^{-\lambda t} = (1.50 \text{ } \mu\text{Ci}) \text{ } e^{-(0.04621 \text{ h}^{-1})(24.0 \text{ h})} = 0.495 \text{ } \mu\text{Ci}.$$

EVALUATE: In 15 h, dN/dt would be 0.75 μCi, and our result in part (d) is less than that, so it is reasonable.

VP43.9.2. IDENTIFY: This problem deals with radioactive decay. The half-life is $T_{1/2} = 35.0$ h and the initial activity is $dN/dt = -0.514$ μCi.

SET UP and EXECUTE: **(a)** $\lambda = (\ln 2)/T_{1/2} = (\ln 2)/(35.0$ h$) = 1.98 \times 10^{-2}$ h$^{-1} = 5.50 \times 10^{-6}$ s^{-1}.

(b) We want the present number N_0 of Nb-95 nuclei.

$$N = N_0 e^{-\lambda t}$$

$$dN/dt = -\lambda N_0 e^{-\lambda t} = -\lambda N_0$$

Call $t = 0$ the present instant when $dN/dt = -0.514$ μCi. Solve the above equation for N_0.

$$N_0 = -\frac{1}{\lambda}\frac{dN}{dt} = -\frac{1}{5.50 \times 10^{-6} \text{ s}^{-1}}(-0.514 \text{ } \mu\text{Ci})(3.70 \times 10^{10} \text{ s}^{-1}) = 3.46 \times 10^9.$$

(c) We want the initial number of Nb-95 nuclei. The 3.46×10^9 nuclei from part (b) is 23.8% of the original number of nuclei. So $0.238N_0 = 3.46 \times 10^9$, which gives $N_0 = 1.45 \times 10^{10}$.

(d) We want the time to now. Solve for t and use the results we have found.

$$N = N_0 e^{-\lambda t}$$

$$t = -\frac{\ln(N/N_0)}{\lambda} = \frac{\ln\left(\dfrac{3.46 \times 10^9}{1.45 \times 10^{10}}\right)}{0.0198 \text{ h}^{-1}} = 72.5 \text{ h}.$$

EVALUATE: The number of Mo-95 nuclei present now is $(0.762)(1.45 \times 10^{10}) = 1.11 \times 10^{10}$. The sum of nuclei now is $1.11 \times 10^{10} + 3.46 \times 10^9 = 1.45 \times 10^{10}$, which agrees with our result in (c).

VP43.9.3. **IDENTIFY:** We are dealing with radioactive decay. The initial activity is 0.415 μCi, and 45.0 s later it is 0.121 μCi.

SET UP and EXECUTE: (a) We want the decay constant and half-life. If we call R the activity, we have seen that it decreases exponentially. Use this fact and solve for the decay constant and then use it to find the half-life.

$$R = R_0 e^{-\lambda t}$$

$$\lambda = -\frac{\ln(R/R_0)}{t} = -\frac{\ln\left(\dfrac{0.121\ \mu\text{Ci}}{0.425\ \mu\text{Ci}}\right)}{45.0\ \text{s}} = 0.0279\ \text{s}^{-1}.$$

$$T_{1/2} = (\ln 2)/\lambda = (\ln 2)/(0.279\ \text{s}^{-1}) = 24.8\ \text{s}.$$

(b) We want the number of nuclei in an excited state. These are the nuclei that give off the radiation, so we want to find the number of undecayed nuclei initially and 45.0 s later.
Initially:

$$dN/dt = -\lambda N_0 e^{-\lambda t} = -\lambda N_0$$

$$(-0.425\ \mu\text{Ci})(3.70 \times 10^{10}\ \text{s}^{-1}) = -(0.0279\ \text{s}^{-1})N_0$$

$$N_0 = 5.63 \times 10^5.$$

After 45.0 s: Using the same approach gives $(-0.121\ \mu\text{Ci})(3.70 \times 10^{10}\ \text{s}^{-1}) = -(0.0279\ \text{s}^{-1})N_0$
$N_0 = 1.60 \times 10^5.$

EVALUATE: The number at 45.0 s is less than the initial number, so our result is reasonable.

VP43.9.4. **IDENTIFY:** This problem is about carbon-14 dating.

SET UP: The half-life of C-14 is 5730 y. If we call R the activity level per gram, it follows from previous work that $R = R_0 e^{-\lambda t}$.

EXECUTE: (a) We want the age of the sample. The activity per unit mass that you detect now is 113 decays per 20 min per 0.510 g of carbon. So $R = [(113\ \text{decays})/(20\ \text{min})]/(0.510\ \text{g}) = 0.18464$ Bq/g. Use $R = R_0 e^{-\lambda t}$ with $\lambda = (\ln 2)/T_{1/2}$ and solve for t.

$$t = \frac{\ln(R/R_0)}{\lambda} = -\frac{T_{1/2}\ln(R/R_0)}{\ln 2} = -\frac{(5730\ \text{y})\ln\left(\dfrac{0.18464\ \text{Bq/g}}{0.255\ \text{Bq/g}}\right)}{45.0\ \text{s}} = 2670\ \text{y}.$$

(b) We want the current number of C-14 nuclei. Call $t = 0$ the present time. In your sample, the present activity is $(113\ \text{decays})/(20\ \text{min}) = 0.094167$ Bq. At $t = 0$ (the present time), we have

$$N_0 = \frac{1}{\lambda}\frac{dN}{dt} = -\frac{T_{1/2}}{\ln 2}(-0.094167\ \text{Bq}) = \frac{(5730\ \text{y})(0.094167\ \text{Bq})}{\ln 2} = 2.46 \times 10^{10}.$$

EVALUATE: The age of your sample is less than one half-life, so its activity (0.18464 Bq/g) should be greater than half the original activity (0.255 Bq/g), which is what we have found.

43.5. **IDENTIFY:** The binding energy of the nucleus is the energy of its constituent particles minus the energy of the carbon-12 nucleus.

SET UP: In terms of the masses of the particles involved, the binding energy is
$E_B = (6m_H + 6m_n - m_{C-12})c^2.$

EXECUTE: (a) Using the values from Table 43.2, we get
$E_B = [6(1.007825\ \text{u}) + 6(1.008665\ \text{u}) - 12.000000\ \text{u}](931.5\ \text{MeV/u}) = 92.16\ \text{MeV}.$

(b) The binding energy per nucleon is $(92.16\ \text{MeV})/(12\ \text{nucleons}) = 7.680\ \text{MeV/nucleon}.$

(c) The energy of the C-12 nucleus is $(12.0000\ \text{u})(931.5\ \text{MeV/u}) = 11178\ \text{MeV}.$ Therefore the

percent of the mass that is binding energy is $\dfrac{92.16\ \text{MeV}}{11178\ \text{MeV}} = 0.8245\%.$

EVALUATE: The binding energy of 92.16 MeV binds 12 nucleons. The binding energy per nucleon, rather than just the total binding energy, is a better indicator of the strength with which a nucleus is bound.

43.7. **IDENTIFY:** This problem is about the binding energy of the atomic nucleus.

SET UP and EXECUTE: The target variable is the mass of a neutral Fe-56 atom. Since the atom is neutral, it contains all of its electrons. Use Eq. (43.10) and solve for the mass of Fe-56.

$$_{26}^{56}M = ZM_H + Nm_n - E_B/c^2$$

$$_{26}^{56}M = 26(1.007825\ u) + 30(1.008665\ u) - \left[(8.79\ \text{MeV})(56)/c^2 \right] \left(\frac{1\ u}{931.5\ \text{MeV}/c^2} \right) = 55.935\ u.$$

EVALUATE: If there were no binding energy, the mass would be $26M_H + 30m_n = 56.463$ u. The binding energy makes a detectable difference in the mass of the atom.

43.11. **IDENTIFY:** We are looking at the decay of C-11.

SET UP and EXECUTE: **(a)** In positron decay, a proton changes to a neutron and an electron. So the number of protons decreases from 6 to 5 and the number of neutrons increases from 5 to 6. So the daughter nucleus contains 5 protons and 6 neutrons.
(b) For positron decay to occur, the mass of the parent must be at least two electron masses greater than the mass of the daughter. The daughter nucleus has 5 protons, so it is boron (B). The decay is

$$_{6}^{11}C \rightarrow _{5}^{11}B + \beta^+ + \nu_e.$$

Use Table 43.2 for the mass of boron and $m_e = 0.000548580$ u. The initial mass is 11.011433 u and the final mass is 11.009305 u + 2(0.000548580 u) = 11.01040216 u. The difference in mass is 11.01040216 u − 11.011433 u = −0.001030840 u. The energy released is the energy of the lost mass, which is (0.001030840 u)(931.5 MeV/u) = 0.960 MeV.
EVALUATE: Notice that we need to use *twice* the mass of the positron for positron decay.

43.15. **IDENTIFY:** This problem is about β^+ decay.

SET UP and EXECUTE: For β^+ decay, the mass of the decay products must be at least two electron-masses greater than the mass of the original atom. The decay in this case is

$$_{7}^{12}N \rightarrow _{6}^{12}C + \beta^+ + \nu_e.$$

The target variable is the mass M of the N-12. Using the energy released given in the problem, we have $M = m_C + 2m_e + 16.316$ MeV/c^2. Using Table 43.2 gives
$M = 12.000000$ u + 2(0.000548580 u) + (16.316 MeV)[u/(931.5 MeV)] = 12.018611 u.
EVALUATE: Check: M is greater than the mass of C-12, as it should be.

43.17. **IDENTIFY:** Determine the energy released during tritium decay.

SET UP: In beta decay an electron, e^-, is emitted by the nucleus. The beta decay reaction is $_{1}^{3}H \rightarrow e^- + _{2}^{3}He$. If neutral atom masses are used, $_{1}^{3}H$ includes one electron and $_{2}^{3}He$ includes two electrons. One electron mass cancels and the other electron mass in $_{2}^{3}He$ represents the emitted electron. Or, we can subtract the electron masses and use the nuclear masses. The atomic mass of $_{2}^{3}He$ is 3.016029 u.

EXECUTE: **(a)** The mass of the $_{1}^{3}H$ nucleus is 3.016049 u − 0.000549 u = 3.015500 u. The mass of the $_{2}^{3}He$ nucleus is 3.016029 u − 2(0.000549 u) = 3.014931 u. The nuclear mass of $_{2}^{3}He$ plus the mass of the emitted electron is 3.014931 u + 0.000549 u = 3.015480 u. This is slightly less than the nuclear mass for $_{1}^{3}H$, so the decay is energetically allowed.

(b) The mass decrease in the decay is $3.015500 \, u - 3.015480 \, u = 2.0 \times 10^{-5} \, u$. Note that this can also be calculated as $m(_1^3 H) - m(_2^4 He)$, where atomic masses are used. The energy released is $(2.0 \times 10^{-5} \, u)(931.5 \, MeV/u) = 0.019 \, MeV$. The total kinetic energy of the decay products is 0.019 MeV, or 19 keV.

EVALUATE: The energy is not shared equally by the decay products because they have unequal masses.

43.19. **IDENTIFY and SET UP:** $T_{1/2} = \dfrac{\ln 2}{\lambda}$ The mass of a single nucleus is $124 m_p = 2.07 \times 10^{-25} \, kg$.

$$|dN/dt| = 0.350 \, Ci = 1.30 \times 10^{10} \, Bq, \quad |dN/dt| = \lambda N.$$

EXECUTE: $N = \dfrac{6.13 \times 10^{-3} \, kg}{2.07 \times 10^{-25} \, kg} = 2.96 \times 10^{22}; \quad \lambda = \dfrac{|dN/dt|}{N} = \dfrac{1.30 \times 10^{10} \, Bq}{2.96 \times 10^{22}} = 4.39 \times 10^{-13} \, s^{-1}.$

$$T_{1/2} = \frac{\ln 2}{\lambda} = 1.58 \times 10^{12} \, s = 5.01 \times 10^4 \, y.$$

EVALUATE: Since $T_{1/2}$ is very large, the activity changes very slowly.

43.21. **IDENTIFY:** From the known half-life, we can find the decay constant, the rate of decay, and the activity.

SET UP: $\lambda = \dfrac{\ln 2}{T_{1/2}}$. $T_{1/2} = 4.47 \times 10^9 \, yr = 1.41 \times 10^{17} \, s$. The activity is $\left| \dfrac{dN}{dt} \right| = \lambda N$. The mass of one ^{238}U is approximately $238 m_p$. $1 \, Ci = 3.70 \times 10^{10}$ decays/s.

EXECUTE: **(a)** $\lambda = \dfrac{\ln 2}{1.41 \times 10^{17} \, s} = 4.92 \times 10^{-18} \, s^{-1}.$

(b) $N = \dfrac{|dN/dt|}{\lambda} = \dfrac{3.70 \times 10^{10} \, Bq}{4.92 \times 10^{-18} \, s^{-1}} = 7.52 \times 10^{27}$ nuclei. The mass m of uranium is the number of nuclei times the mass of each one. $m = (7.52 \times 10^{27})(238)(1.67 \times 10^{-27} \, kg) = 2.99 \times 10^3 \, kg.$

(c) $N = \dfrac{10.0 \times 10^{-3} \, kg}{238 m_p} = \dfrac{10.0 \times 10^{-3} \, kg}{238(1.67 \times 10^{-27} \, kg)} = 2.52 \times 10^{22}$ nuclei.

$\left| \dfrac{dN}{dt} \right| = \lambda N = (4.92 \times 10^{-18} \, s^{-1})(2.52 \times 10^{22}) = 1.24 \times 10^5$ decays/s.

EVALUATE: Because ^{238}U has a very long half-life, it requires a large amount (about 3000 kg) to have an activity of a 1.0 Ci.

43.25. **IDENTIFY:** This problem is about radioactive decay.

SET UP: We are working with the half-life of the isotope, so it is convenient to express the decay in terms of base 2 when needed. $N = N_0 e^{-\lambda t} = N_0 2^{-t/T_{1/2}}$

EXECUTE: **(a)** We want to know the number of nuclei initially present. We know the initial rate of decay, which is dN/dt, so

$$dN/dt = d(N_0 e^{-\lambda t})/dt = -\lambda N_0 e^{-\lambda t}.$$

At $t = 0$, $dN/dt = -8.0 \times 10^{16}$ Bq. Solving for N_0 gives

$$N_0 = \frac{1}{\lambda} \frac{dN}{dt} = -\frac{T_{1/2}}{0.693} \frac{dN}{dt} = -\frac{(64.0)(3600 \, s)}{0.693} \left(-8.0 \times 10^{16} \, Bq \right) = 2.7 \times 10^{22} \text{ nuclei.}$$

(b) We want the number of nuclei remaining at the end of 12.0 days. Use base 2 for simplicity.
$$N = N_0 2^{-t/T_{1/2}} = (2.66\times10^{22})2^{-(12.0)(24.\text{h})(64.0\text{h})} = 1.2\times10^{21}\,\text{nuclei}.$$
EVALUATE: Note that 12.0 days is 4.5 half-lives, so $N/N_0 = 2^{-4.5}$.

43.27. IDENTIFY: Apply $A = A_0 e^{-\lambda t}$ and $\lambda = \ln 2/T_{1/2}$.

SET UP: $\ln e^x = x$.

EXECUTE: $A = A_0 e^{-\lambda t} = A_0 e^{-t(\ln 2)/T_{1/2}}.\quad -\dfrac{(\ln 2)t}{T_{1/2}} = \ln(A/A_0).$

$$T_{1/2} = -\frac{(\ln 2)t}{\ln(A/A_0)} = -\frac{(\ln 2)(4.00\ \text{days})}{\ln(3091/8318)} = 2.80\ \text{days}.$$

EVALUATE: The activity has decreased by more than half and the elapsed time is more than one half-life.

43.29. IDENTIFY and SET UP: Apply $|dN/dt| = \lambda N$ with $\lambda = \ln 2/T_{1/2}$. In one half-life, one half of the nuclei decay.

EXECUTE: **(a)** $\left|\dfrac{dN}{dt}\right| = 7.56\times10^{11}\ \text{Bq} = 7.56\times10^{11}\ \text{decays/s}.$

$$\lambda = \frac{0.693}{T_{1/2}} = \frac{0.693}{(30.8\ \text{min})(60\ \text{s/min})} = 3.75\times10^{-4}\ \text{s}^{-1}.$$

$$N_0 = \frac{1}{\lambda}\left|\frac{dN}{dt}\right| = \frac{7.56\times10^{11}\ \text{decay/s}}{3.75\times10^{-4}\ \text{s}^{-1}} = 2.02\times10^{15}\ \text{nuclei}.$$

(b) The number of nuclei left after one half-life is $\dfrac{N_0}{2} = 1.01\times10^{15}$ nuclei, and the activity is half:

$$\left|\frac{dN}{dt}\right| = 3.78\times10^{11}\ \text{decays/s}.$$

(c) After three half-lives (92.4 minutes) there is an eighth of the original amount, so $N = 2.53\times10^{14}$ nuclei, and an eighth of the activity: $\left|\dfrac{dN}{dt}\right| = 9.45\times10^{10}$ decays/s.

EVALUATE: Since the activity is proportional to the number of radioactive nuclei that are present, the activity is halved in one half-life.

43.33. IDENTIFY and SET UP: For x rays $RBE = 1$ and the equivalent dose equals the absorbed dose.

EXECUTE: **(a)** 175 krad $= 175$ krem $= 1.75$ kGy $= 1.75$ kSv. $(1.75\times10^3\ \text{J/kg})(0.220\ \text{kg}) = 385\ \text{J}.$
(b) 175 krad $= 1.75$ kGy; $(1.50)(175\ \text{krad}) = 262.5$ krem $= 2.625$ kSv. The energy deposited would be 385 J, the same as in (a).
EVALUATE: The energy required to raise the temperature of 0.220 kg of water 1 C° is 922 J, and 385 J is less than this. The energy deposited corresponds to a very small amount of heating.

43.35. IDENTIFY: This problem looks at the biological effects of radiation.

SET UP: $N = N_0 e^{-\lambda t}$, $\lambda = 0.693/T_{1/2} = 0.693/(29\ \text{y}) = 0.0239\ \text{y}^{-1}$.
EXECUTE: **(a)** We want the absorbed dose during one year from 1.0 μg of Sr-90. First find the number of decays ΔN that occur during one year. Each decay releases 1.1 MeV of energy. The mass of a Sr-90 atom is approximately $38m_p + 52m_n + 38m_e = 1.5067 \times 10^{-25}$ kg. If N_0 is the number of Sr-90 atoms in the 1.0-μg sample, then $(1.5067 \times 10^{-25}$ kg$)N_0 = 1.0\ \mu$g, which gives $N_0 = 6.637 \times$

10^{15} atoms. At the end of one year, the number of atoms N that are left will be $N = N_0 e^{-\lambda(1\,y)}$. The number of decays during the year is $N_0 - N = N_0 - N_0 e^{-\lambda(1\,y)} = (6.637 \times 10^{15})[1 - e^{-(0.0239/y)(1\,y)}] = 1.567 \times 10^{14}$ decays. The total energy E absorbed by these decays is $E = (1.567 \times 10^{14})(1.1\ \text{MeV}) = 27.59\ \text{J}$. This energy is delivered to 50 kg of body tissue, so the absorbed dose is $(27.59\ \text{J})/(50\ \text{kg}) = 0.55\ \text{J/kg} = 0.55\ \text{Gy}$. Since 1 rad = 0.01 Gy, this dose is also 55 rad.

(b) We want the equivalent dose. Equivalent dose (in Sv) = RBE × absorbed dose (in Gy). From Table 43.3, the RBE for gamma rays is 1, and the RBE for electrons is given in the problem as 1.0. So the maximum equivalent dose is $(1.0)(0.55\ \text{Gy}) = 0.55\ \text{Sv}$. Since 1 rem = 0.01 Sv, we can also say that the equivalent dose is 55 rem.

EVALUATE: The longer the exposure lasts, the greater the dose because more decays take place.

43.39. **(a) IDENTIFY and SET UP:** Determine X by balancing the charge and the nucleon number on the two sides of the reaction equation.

EXECUTE: X must have $A = +2 + 9 - 4 = 7$ and $Z = +1 + 4 - 2 = 3$. Thus X is $^{7}_{3}\text{Li}$ and the reaction is $^{2}_{1}\text{H} + ^{9}_{4}\text{Be} = ^{7}_{3}\text{Li} + ^{4}_{2}\text{He}$.

(b) IDENTIFY and SET UP: Calculate the mass decrease and find its energy equivalent.

EXECUTE: If we use the neutral atom masses then there are the same number of electrons (five) in the reactants as in the products. Their masses cancel, so we get the same mass defect whether we use nuclear masses or neutral atom masses. The neutral atoms masses are given in Table 43.2. 1 u is equivalent to 931.5 MeV.

$$^{2}_{1}\text{H} + ^{9}_{4}\text{Be} \text{ has mass } 2.014102\ \text{u} + 9.012182\ \text{u} = 11.26284\ \text{u}.$$

$$^{7}_{3}\text{Li} + ^{4}_{2}\text{He} \text{ has mass } 7.016005\ \text{u} + 4.002603\ \text{u} = 11.018608\ \text{u}.$$

The mass decrease is $11.026284\ \text{u} - 11.018608\ \text{u} = 0.007676\ \text{u}$.

This corresponds to an energy release of $(0.007676\ \text{u})(931.5\ \text{MeV/1 u}) = 7.150\ \text{MeV}$.

(c) IDENTIFY and SET UP: Estimate the threshold energy by calculating the Coulomb potential energy when the $^{2}_{1}\text{H}$ and $^{9}_{4}\text{Be}$ nuclei just touch. Obtain the nuclear radii from $R = R_0 A^{1/3}$.

EXECUTE: The radius R_{Be} of the $^{9}_{4}\text{Be}$ nucleus is $R_{\text{Be}} = (1.2 \times 10^{-15}\ \text{m})(9)^{1/3} = 2.5 \times 10^{-15}\ \text{m}$.

The radius R_{H} of the $^{2}_{1}\text{H}$ nucleus is $R_{\text{H}} = (1.2 \times 10^{-15}\ \text{m})(2)^{1/3} = 1.5 \times 10^{-15}\ \text{m}$.

The nuclei touch when their center-to-center separation is

$R = R_{\text{Be}} + R_{\text{H}} = 4.0 \times 10^{-15}\ \text{m}$.

The Coulomb potential energy of the two reactant nuclei at this separation is

$$U = \frac{1}{4\pi\epsilon_0} \frac{q_1 q_2}{r} = \frac{1}{4\pi\epsilon_0} \frac{e(4e)}{r}.$$

$$U = (8.988 \times 10^9\ \text{N} \cdot \text{m}^2/\text{C}^2) \frac{4(1.602 \times 10^{-19}\ \text{C})^2}{(4.0 \times 10^{-15}\ \text{m})(1.602 \times 10^{-19}\ \text{J/eV})} = 1.4\ \text{MeV}.$$

This is an estimate of the threshold energy for this reaction.

EVALUATE: The reaction releases energy but the total initial kinetic energy of the reactants must be 1.4 MeV in order for the reacting nuclei to get close enough to each other for the reaction to occur. The nuclear force is strong but is very short-range.

43.41. **IDENTIFY** and **SET UP:** The energy released is the energy equivalent of the mass decrease. 1 u is equivalent to 931.5 MeV. The mass of one ^{235}U nucleus is $235m_p$.

EXECUTE: **(a)** $^{235}_{92}U + {}^1_0n \rightarrow {}^{144}_{56}Ba + {}^{89}_{36}Kr + 3{}^1_0n$. We can use atomic masses since the same number of electrons are included on each side of the reaction equation and the electron masses cancel. The mass decrease is $\Delta M = m({}^{235}_{92}U) + m({}^1_0n) - [m({}^{144}_{56}Ba) + m({}^{89}_{36}Kr) + 3m({}^1_0n)]$,
$\Delta M = 235.043930 \text{ u} + 1.0086649 \text{ u} - 143.922953 \text{ u} - 88.917631 \text{ u} - 3(1.0086649 \text{ u})$,
$\Delta M = 0.1860$ u. The energy released is $(0.1860 \text{ u})(931.5 \text{ MeV/u}) = 173.3$ MeV.

(b) The number of ^{235}U nuclei in 1.00 g is $\dfrac{1.00 \times 10^{-3} \text{ kg}}{235m_p} = 2.55 \times 10^{21}$. The energy released per gram is $(173.3 \text{ MeV/nucleus})(2.55 \times 10^{21} \text{ nuclei/g}) = 4.42 \times 10^{23}$ MeV/g.

EVALUATE: The energy released is 7.1×10^{10} J/kg. This is much larger than typical heats of combustion, which are about 5×10^4 J/kg.

43.45. **IDENTIFY** and **SET UP:** $m = \rho V$. $1 \text{ gal} = 3.788 \text{ L} = 3.788 \times 10^{-3} \text{ m}^3$. The mass of a ^{235}U nucleus is $235m_p$. $1 \text{ MeV} = 1.60 \times 10^{-13}$ J.

EXECUTE: **(a)** For 1 gallon, $m = \rho V = (737 \text{ kg/m}^3)(3.788 \times 10^{-3} \text{ m}^3) = 2.79 \text{ kg} = 2.79 \times 10^3$ g.
$$\frac{1.3 \times 10^8 \text{ J/gal}}{2.79 \times 10^3 \text{ g/gal}} = 4.7 \times 10^4 \text{ J/g}.$$

(b) 1 g contains $\dfrac{1.00 \times 10^{-3} \text{ kg}}{235m_p} = 2.55 \times 10^{21}$ nuclei.
$(200 \text{ MeV/nucleus})(1.60 \times 10^{-13} \text{ J/MeV})(2.55 \times 10^{21} \text{ nuclei}) = 8.2 \times 10^{10}$ J/g.

(c) A mass of $6m_p$ produces 26.7 MeV.
$$\frac{(26.7 \text{ MeV})(1.60 \times 10^{-13} \text{ J/MeV})}{6m_p} = 4.26 \times 10^{14} \text{ J/kg} = 4.26 \times 10^{11} \text{ J/g}.$$

(d) The total energy available would be $(1.99 \times 10^{30} \text{ kg})(4.7 \times 10^7 \text{ J/kg}) = 9.4 \times 10^{37}$ J.
$$\text{Power} = \frac{\text{energy}}{t} \text{ so } t = \frac{\text{energy}}{\text{power}} = \frac{9.4 \times 10^{37} \text{ J}}{3.86 \times 10^{26} \text{ W}} = 2.4 \times 10^{11} \text{ s} = 7600 \text{ y}.$$

EVALUATE: If the mass of the sun were all proton fuel, it would contain enough fuel to last
$$(7600 \text{ y})\left(\frac{4.3 \times 10^{11} \text{ J/g}}{4.7 \times 10^4 \text{ J/g}}\right) = 7.0 \times 10^{10} \text{ y}.$$

43.47. **IDENTIFY:** The minimum energy to remove a proton or a neutron from the nucleus is equal to the energy difference between the two states of the nucleus, before and after removal.

(a) **SET UP:** $^{17}_8O = {}^1_0n + {}^{16}_8O$. $\Delta m = m({}^1_0n) + m({}^{16}_8O) - m({}^{17}_8O)$. The electron masses cancel when neutral atom masses are used.
EXECUTE: $\Delta m = 1.008665 \text{ u} + 15.994915 \text{ u} - 16.999132 \text{ u} = 0.004448$ u. The energy equivalent of this mass increase is $(0.004448 \text{ u})(931.5 \text{ MeV/u}) = 4.14$ MeV.

(b) SET UP and EXECUTE: Following the same procedure as in part (a) gives

$$\Delta M = 8M_H + 9M_n - {}^{17}_8 M = 8(1.007825 \text{ u}) + 9(1.008665 \text{ u}) - 16.999132 \text{ u} = 0.1415 \text{ u}.$$

$$E_B = (0.1415 \text{ u})(931.5 \text{ MeV/u}) = 131.8 \text{ MeV}. \quad \frac{E_B}{A} = 7.75 \text{ MeV/nucleon}.$$

EVALUATE: The neutron removal energy is about half the binding energy per nucleon.

43.51. **(a) IDENTIFY and SET UP:** The heavier nucleus will decay into the lighter one.

EXECUTE: ${}^{25}_{13}$Al will decay into ${}^{25}_{12}$Mg.

(b) IDENTIFY and SET UP: Determine the emitted particle by balancing A and Z in the decay reaction.

EXECUTE: This gives ${}^{25}_{13}$Al $\rightarrow {}^{25}_{12}$Mg $+ {}^{0}_{+1}$e. The emitted particle must have charge $+e$ and its nucleon number must be zero. Therefore, it is a β^+ particle, a positron.

(c) IDENTIFY and SET UP: Calculate the energy defect ΔM for the reaction and find the energy equivalent of ΔM. Use the nuclear masses for ${}^{25}_{13}$Al and ${}^{25}_{12}$Mg, to avoid confusion in including the correct number of electrons if neutral atom masses are used.

EXECUTE: The nuclear mass for ${}^{25}_{13}$Al is

$$M_{\text{nuc}} ({}^{25}_{13}\text{Al}) = 24.990428 \text{ u} - 13(0.000548580 \text{ u}) = 24.983296 \text{ u}.$$

The nuclear mass for ${}^{25}_{12}$Mg is $M_{\text{nuc}} ({}^{25}_{12}\text{Mg}) = 24.985837 \text{ u} - 12(0.000548580 \text{ u}) = 24.979254 \text{ u}.$
The mass defect for the reaction is

$$\Delta M = M_{\text{nuc}} ({}^{25}_{13}\text{Al}) - M_{\text{nuc}} ({}^{25}_{12}\text{Mg}) - M({}^{0}_{+1}\text{e}) = 24.983296 \text{ u} - 24.979254 \text{ u} - 0.00054858 \text{ u}$$
$$= 0.003493 \text{ u}.$$

$$Q = (\Delta M)c^2 = 0.003493 \text{ u}(931.5 \text{ MeV/1 u}) = 3.254 \text{ MeV}.$$

EVALUATE: The mass decreases in the decay and energy is released. Note:
${}^{25}_{13}$Al can also decay into ${}^{25}_{12}$Mg by the electron capture: ${}^{25}_{13}$Al $+ {}^{0}_{-1}$e $\rightarrow {}^{25}_{12}$Mg. The ${}^{0}_{-1}$ electron in the reaction is an orbital electron in the neutral ${}^{25}_{13}$Al atom. The mass defect can be calculated using the nuclear masses:

$$\Delta M = M_{\text{nuc}} ({}^{25}_{13}\text{Al}) + M({}^{0}_{-1}\text{e}) - M_{\text{nuc}} ({}^{25}_{12}\text{Mg}) = 24.983296 \text{ u} + 0.00054858 \text{ u} - 24.979254 \text{ u}$$
$$= 0.004591 \text{ u}.$$

$$Q = (\Delta M) c^2 = (0.004591 \text{ u})(931.5 \text{ MeV/1 u}) = 4.277 \text{ MeV}.$$ The mass decreases in the decay and energy is released.

43.53. **IDENTIFY and SET UP:** The amount of kinetic energy released is the energy equivalent of the mass change in the decay. $m_e = 0.0005486$ u and the atomic mass of ${}^{14}_7$N is 14.003074 u. The energy equivalent of 1 u is 931.5 MeV. ^{14}C has a half-life of $T_{1/2} = 5730$ y $= 1.81 \times 10^{11}$ s. The RBE for an electron is 1.0.

EXECUTE: **(a)** ${}^{14}_6$C $\rightarrow$ e$^-$ $+ {}^{14}_7$N $+ \bar{\nu}_e$.

(b) The mass decrease is $\Delta M = m({}^{14}_6$C$) - [m_e + m({}^{14}_7$N$)]$. Use nuclear masses, to avoid difficulty in accounting for atomic electrons. The nuclear mass of ${}^{14}_6$C is 14.003242 u $- 6m_e = 13.999950$ u. The nuclear mass of ${}^{14}_7$N is 14.003074 u $- 7m_e = 13.999234$ u.

$\Delta M = 13.999950$ u $- 13.999234$ u $- 0.000549$ u $= 1.67 \times 10^{-4}$ u. The energy equivalent of ΔM is 0.156 MeV.

(c) The mass of carbon is $(0.18)(75 \text{ kg}) = 13.5$ kg. From Example 43.9, the activity due to 1 g of carbon in a living organism is 0.255 Bq. The number of decay/s due to 13.5 kg of carbon is
$$(13.5 \times 10^3 \text{ g})(0.255 \text{ Bq/g}) = 3.4 \times 10^3 \text{ decays/s.}$$

(d) Each decay releases 0.156 MeV so 3.4×10^3 decays/s releases 530 MeV/s $= 8.5 \times 10^{-11}$ J/s.

(e) The total energy absorbed in 1 year is $(8.5 \times 10^{-11} \text{ J/s})(3.156 \times 10^7 \text{ s}) = 2.7 \times 10^{-3}$ J. The absorbed dose is $\dfrac{2.7 \times 10^{-3} \text{ J}}{75 \text{ kg}} = 3.6 \times 10^{-5}$ J/kg $= 36$ μGy $= 3.6$ mrad. With RBE $= 1.0$, the equivalent dose is 36 μSv $= 3.6$ mrem.

EVALUATE: Section 43.5 says that background radiation exposure is about 1.0 mSv per year. The radiation dose calculated in this problem is much less than this.

43.55. **IDENTIFY and SET UP:** The mass defect is E_B/c^2.

EXECUTE: $m_{^{11}_{6}\text{C}} - m_{^{11}_{5}\text{B}} - 2m_e = 1.03 \times 10^{-3}$ u. Decay is energetically possible.

EVALUATE: The energy released in the decay is $(1.03 \times 10^{-3} \text{ u})(931.5 \text{ MeV/u}) = 0.959$ MeV.

43.59. **IDENTIFY:** Use $N = N_0 e^{-\lambda t}$ to relate the initial number of radioactive nuclei, N_0, to the number, N, left after time t.

SET UP: We have to be careful; after ^{87}Rb has undergone radioactive decay it is no longer a rubidium atom. Let N_{85} be the number of ^{85}Rb atoms; this number doesn't change. Let N_0 be the number of ^{87}Rb atoms on earth when the solar system was formed. Let N be the present number of ^{87}Rb atoms.

EXECUTE: The present measurements say that $0.2783 = N/(N + N_{85})$.
$(N + N_{85})(0.2783) = N$, so $N = 0.3856 N_{85}$. The percentage we are asked to calculate is $N_0/(N_0 + N_{85})$.

N and N_0 are related by $N = N_0 e^{-\lambda t}$ so $N_0 = e^{+\lambda t} N$.

Thus $\dfrac{N_0}{N_0 + N_{85}} = \dfrac{N e^{\lambda t}}{N e^{\lambda t} + N_{85}} = \dfrac{(0.3855 e^{\lambda t}) N_{85}}{(0.3856 e^{\lambda t}) N_{85} + N_{85}} = \dfrac{0.3856 e^{\lambda t}}{0.3856 e^{\lambda t} + 1}$.

$$t = 4.6 \times 10^9 \text{ y}; \quad \lambda = \frac{0.693}{T_{1/2}} = \frac{0.693}{4.75 \times 10^{10} \text{ y}} = 1.459 \times 10^{-11} \text{ y}^{-1}.$$

$$e^{\lambda t} = e^{(1.459 \times 10^{-11} \text{ y}^{-1})(4.6 \times 10^9 \text{ y})} = e^{0.06711} = 1.0694.$$

Thus $\dfrac{N_0}{N_0 + N_{85}} = \dfrac{(0.3856)(1.0694)}{(0.3856)(1.0694) + 1} = 29.2\%$.

EVALUATE: The half-life for ^{87}Rb is a factor of 10 larger than the age of the solar system, so only a small fraction of the ^{87}Rb nuclei initially present have decayed; the percentage of rubidium atoms that are radioactive is only a bit less now than it was when the solar system was formed.

43.63. **IDENTIFY:** This problem investigates the p-p II fusion chain in the sun.

SET UP: Follow the directions in each part and use Table 43.2 for nuclide masses. The third step of the chain is

$$\,^3_2\text{He} + \,^3_2\text{He} \rightarrow \,^7_4\text{Be} + \gamma.$$

EXECUTE: **(a)** We want the energy E_γ of the photon. Realize that $E_\gamma = Q$. From Table 43.2 we have:

Mass of He-3 = 3.016029 u
Mass of He-4 = 4.002603 u
Mass of Be-7 = 7.016930 u

The nuclide masses include the electrons, but since we'll be subtracting their effect subtracts out. Therefore $E_\gamma = Q = (3.016029 \text{ u} + 4.002603 \text{ u} - 7.016930 \text{ u})(931.5 \text{ MeV/u}) = 1.59 \text{ MeV}$.

(b) We want the energy E_ν of the neutrino in the following electron-capture of Be-7.

$$\,^7_4\text{Be} + e^- \rightarrow \,^7_3\text{Li} + \nu_e.$$

Using the given mass of Li-7 and the Be-7 mass from Table 43.2, we have

$$E_\nu = (7.016930 \text{ u} - 7.016003 \text{ u})(931.5 \text{ MeV/u}) = 0.864 \text{ MeV}.$$

(c) We want the energy Q that is released. Using the given reaction, we have

$$Q = [1.007825 \text{ u} + 7.016003 \text{ u} - 2(4.002603 \text{ u})](931.5 \text{ MeV/u}) = 17.35 \text{ MeV}.$$

(d) We want the speed v_α of the alpha particles. The alpha particles have equal speeds, so for each one $K_\alpha = (17.35 \text{ MeV})/2 = 8.675 \text{ MeV}$. Use the relativistic kinetic energy $K = mc^2(\gamma - 1)$. Solving for γ gives

$$\gamma = 1 + K_\alpha / m_\alpha c^2 = 1 + \frac{8.675 \text{ MeV}}{(4.002603 \text{ u})(931.5 \text{ MeV/u})} = 1.0023267.$$

Using this result gives

$$v = c\sqrt{1 - 1/\gamma^2} = 0.0681c.$$

(e) We want the total energy. The overall p-p II chain can be summarized as

$$4\,^1_1\text{H} + \,^4_2\text{He} \rightarrow 2\,^4_2\text{He} + 2\nu_e + \gamma.$$

Using masses from Table 43.2 gives $Q = [4(1.007825 \text{ u}) - 4.002603 \text{ u}](931.5 \text{ MeV/u}) = 26.7 \text{ MeV}$.

(f) The p-p I releases 26.73 MeV, so $E_{\text{II}}/E_\text{I} = (26.7 \text{ MeV})/(26.73 \text{ MeV}) = 99.9\%$. They are about the same.

(g) We want E_ν/E_{tot}. Using the results from (b) and (f) gives $E_\nu/E_\text{I} = (0.864 \text{ MeV})/(26.7 \text{ MeV}) = 3.2\%$.

EVALUATE: The amount of energy per reaction is the same for p-p I and p-p II, but p-p II occurs less frequently so it contributes less to the sun's total energy production.

43.65. **IDENTIFY and SET UP:** The number of radioactive nuclei left after time t is given by $N = N_0 e^{-\lambda t}$. The problem says $N/N_0 = 0.29$; solve for t.

EXECUTE: $0.29 = e^{-\lambda t}$ so $\ln(0.29) = -\lambda t$ and $t = -\ln(0.29)/\lambda$. Example 43.9 gives $\lambda = 1.209\times10^{-4}$ y^{-1} for ^{14}C. Thus $t = \dfrac{-\ln(0.29)}{1.209\times10^{-4} \text{ y}} = 1.0\times10^4$ y.

EVALUATE: The half-life of ^{14}C is 5730 y, so our calculated t is about 1.75 half-lives, so the fraction remaining is around $\left(\frac{1}{2}\right)^{1.75} = 0.30$.

43.67. **(a) IDENTIFY** and **SET UP:** Use $R = R_0 A^{1/3}$ to calculate the radius R of a $^2_1 H$ nucleus. Calculate the Coulomb potential energy $U = \dfrac{1}{4\pi\epsilon_0}\dfrac{q_1 q_2}{r^2}$ of the two nuclei when they just touch.

EXECUTE: The radius of $^2_1 H$ is $R = (1.2\times10^{-15}\text{ m})(2)^{1/3} = 1.51\times10^{-15}$ m. The barrier energy is the Coulomb potential energy of two $^2_1 H$ nuclei with their centers separated by twice this distance:

$$U = \frac{1}{4\pi\epsilon_0}\frac{e^2}{r} = (8.988\times10^9\text{ N}\cdot\text{m}^2/\text{C}^2)\frac{(1.602\times10^{-19}\text{ C})^2}{2(1.51\times10^{-15}\text{ m})} = 7.64\times10^{-14}\text{ J} = 0.48\text{ MeV}.$$

(b) IDENTIFY and **SET UP:** Find the energy equivalent of the mass decrease.

EXECUTE: $^2_1 H + {}^2_1 H \rightarrow {}^3_2 He + {}^1_0 n$

If we use neutral atom masses there are two electrons on each side of the reaction equation, so their masses cancel. The neutral atom masses are given in Table 43.2.

$$^2_1 H + {}^2_1 H \text{ has mass } 2(2.014102\text{ u}) = 4.028204\text{ u}$$

$$^3_2 He + {}^1_0 n \text{ has mass } 3.016029\text{ u} + 1.008665\text{ u} = 4.024694\text{ u}$$

The mass decrease is $4.028204\text{ u} - 4.024694\text{ u} = 3.510\times10^{-3}$ u. This corresponds to a liberated energy of

$$(3.510\times10^{-3}\text{ u})(931.5\text{ MeV/u}) = 3.270\text{ MeV, or }(3.270\times10^6\text{ eV})(1.602\times10^{-19}\text{ J/eV})$$

$$= 5.239\times10^{-13}\text{ J}.$$

(c) IDENTIFY and **SET UP:** We know the energy released when two $^2_1 H$ nuclei fuse. Find the number of reactions obtained with one mole of $^2_1 H$.

EXECUTE: Each reaction takes two $^2_1 H$ nuclei. Each mole of D_2 has 6.022×10^{23} molecules, so 6.022×10^{23} pairs of atoms. The energy liberated when one mole of deuterium undergoes fusion is $(6.022\times10^{23})(5.239\times10^{-13}\text{ J}) = 3.155\times10^{11}$ J/mol.

EVALUATE: The energy liberated per mole is more than a million times larger than from chemical combustion of one mole of hydrogen gas.

43.69. **IDENTIFY:** Apply $\left|\dfrac{dN}{dt}\right| = \lambda N_0 e^{-\lambda t}$, with $\lambda = \dfrac{\ln 2}{T_{1/2}}$.

SET UP: $\ln|dN/dt| = \ln\lambda N_0 - \lambda t$.

EXECUTE: **(a)** A least-squares fit to log of the activity *vs.* time gives a slope of magnitude

$$\lambda = 0.5995\text{ h}^{-1}, \text{ for a half-life of } \frac{\ln 2}{\lambda} = 1.16\text{ h}.$$

(b) The initial activity is $N_0\lambda$, and this gives $N_0 = \dfrac{(2.00\times10^4\text{ Bq})}{(0.5995\text{ hr}^{-1})(1\text{ hr}/3600\text{ s})} = 1.20\times10^8.$

(c) $N = N_0 e^{-\lambda t} = 1.81\times10^6.$

EVALUATE: The activity decreases by about $\frac{1}{2}$ in the first hour, so the half-life is about 1 hour.

43.71. **IDENTIFY:** Apply $A = A_0 e^{-\lambda t}$, where A is the activity and $\lambda = (\ln 2)/T_{1/2}$. This equation can be written as $A = A_0 2^{-(t/T_{1/2})}$. The activity of the engine oil is proportional to the mass worn from the piston rings.

SET UP: $1 \text{ Ci} = 3.7 \times 10^{10} \text{ Bq}$.

EXECUTE: The activity of the original iron, after 1000 hours of operation, would be $(9.4 \times 10^{-6} \text{ Ci})(3.7 \times 10^{10} \text{ Bq/Ci}) 2^{-(1000 \text{ h})/[(45 \text{ d})(24 \text{ h/d})]} = 1.8306 \times 10^5 \text{ Bq}$. The activity of the oil is 84 Bq, or 4.5886×10^{-4} of the total iron activity, and this must be the fraction of the mass worn, or mass of 4.59×10^{-2} g. The rate at which the piston rings lost their mass is then 4.59×10^{-5} g/h.

EVALUATE: This method is very sensitive and can measure very small amounts of wear.

STUDY GUIDE FOR PARTICLE PHYSICS AND COSMOLOGY

Summary

In this chapter, we look at the formation of the universe through an investigation of particle physics. We'll examine the four fundamental forces through which particles interact. We investigate the rules that govern these interactions and their influence on the universe as we know it. We'll go on to consider the similarities among these forces and how they can be joined, or unified, into fewer forces. We'll then look back and see how the universe has evolved since the Big Bang—the massive explosion that generated the universe.

Objectives

After studying this chapter, you'll understand

- Key fundamental subatomic particles and how they were discovered.
- The four fundamental interactions between particles.
- The rules that govern the four interactions and how to apply them.
- How protons, neutrons, and other particles are built from quarks and gluons.
- How all the particles and interactions fit into the standard model.
- Evidence of the Big Bang and the history of the universe since the Big Bang.

From Chapter 44 of Student's Study Guide to accompany *University Physics with Modern Physics, Volume 3,* Fifteenth Edition.
Hugh D. Young and Roger A. Freedman. Copyright © 2020 by Pearson Education, Inc. All rights reserved.

Concepts and Equations

Term	Description
Fundamental Particles	Protons, neutrons, and other hadrons are made of quarks with fractional charge. Some hadrons, such as protons and neutrons, are made of three quarks. Other hadrons, such as mesons, are made of two quarks. All massive particles (protons, neutrons, electrons, etc.) have antiparticles—particles with the same characteristics, but opposite charge.
	Particles serve as mediators of the fundamental interactions. The photon is the mediator for the electromagnetic force, the pion for the nuclear force.
Particle Accelerators and Detectors	Cyclotrons, synchrotrons, and linear accelerators are used to accelerate particles to high energies so that physicists can study their fundamental interactions. Higher energy accelerators allow scientists to probe deeper into the nucleus. Particle detectors are used to identify reaction products and measure their energies and momenta.
Fundamental Interactions	Four distinct types of interactions or forces are known to exist in the universe: the gravitational interaction, the electromagnetic interaction, the strong interaction, and the weak interaction. Particles are described in terms of their interactions. Conservation laws apply to the certain interactions, but not others.
	Physicists are trying to unify all four forces in a single interaction theory. They have successfully combined the last three interactions into one theory and are testing their predictions. They are currently working on adding gravitation to the other interactions to form a "theory of everything."
Expanding Universe	Hubble's law shows that galaxies are receding from each other and the universe is expanding. Observations show that the rate of expansion is increasing, due to the presence of dark energy, which makes up 74% of the energy in the universe. A total of 4% of the universe is made of ordinary matter that we have studied in this course. The remaining 22% of the universe is dark matter. Both dark matter and dark energy are poorly understood and are the subject of intense study by physicists.
Cosmology	Particle physics leads to an understanding of the development of the universe, or the study of cosmology. The universe began with the Big Bang, in which all the matter and energy of the universe was compressed into a single point. Since the Big Bang, the universe has been cooling and expanding for 14 billion years.

Key Concept 1: In the type of particle accelerator called a cyclotron, particles follow a circular path in a plane perpendicular to a uniform magnetic field. Every one-half orbit the particles pass through a potential difference that increases their kinetic energy and the radius of their circular path. The maximum radius of the cyclotron path determines the maximum kinetic energy attainable.

Key Concept 2: The available energy E_a in a particle collision is the energy available to produce the particles that are present after the collision. This equals the total energy of the colliding particles in the center-of-momentum system, a frame in which the total momentum of the particles is zero. Given E_a, you can calculate the required kinetic energy in a different frame in which one of the particles is initially at rest.

Key Concept 3: In experiments in which a fast-moving particle strikes a second target particle that is initially stationary, a large increase in the energy of the moving particle produces only a relatively small increase in the available energy.

Key Concept 4: There are three types of lepton number: electron lepton number L_e, muon lepton number L_μ, and tau lepton number L_τ. Each of these is separately conserved in particle interactions; no violations of these conservation laws have ever been observed.

Key Concept 5: Baryon number, like the three lepton numbers, is always conserved in particle interactions.

Key Concept 6: To calculate the minimum energy required to produce a new particle, you must first consider the conservation laws associated with the particle. These laws will tell you which other particles may have to be produced along with the desired one. Once you have done this, you can determine how much available energy is required.

Key Concept 7: All baryons are composed of three quarks, and all antibaryons are composed of three antiquarks. For a baryon or antibaryon, the total electric charge equals the sum of the charges of its constituent quarks or antiquarks; the same is true of the total baryon number and total strangeness.

Key Concept 8: The redshift of light from distant galaxies shows that they are moving away from us. The recession speed can be calculated from the ratio of the measured wavelength λ_0 of light from the galaxy to the wavelength λ_S measured in the rest frame of the galaxy.

Key Concept 9: The Hubble law is the observation that the speed at which a distant galaxy moves away from us is directly proportional to its distance from us. The speed of recession can be determined from the galaxy's redshift, and the distance can then be calculated from the Hubble law.

Key Concept 10: The average kinetic energy of particles in a gas is directly proportional to the absolute temperature T. At extremely high temperatures above 10^{10} K, this average kinetic energy can be comparable to the rest energies of subatomic particles.

Key Concept 11: The first atomic nuclei formed about three and a half minutes after the Big Bang. The present-day ratio of hydrogen nuclei to helium nuclei in the universe was determined by the ratio of protons to neutrons at that moment in the history of the universe.

Key Concept 12: The present-day temperature of the cosmic background radiation tells us how much the universe has expanded since the time 380,000 years after the Big Bang when the first atoms formed and the universe became transparent.

Conceptual Questions

1: Fundamental forces

Which of the four fundamental forces influence electrons, neutrinos, and protons?

IDENTIFY, SET UP, AND EXECUTE Electrons are affected by the electromagnetic force, since they have electric charge; the weak force, since they have weak charge; and the gravitational force, since they have mass (i.e., "gravitational charge").

Neutrinos have no mass or electric charge and so are affected only by the weak force. Protons are affected by all four forces, since they have mass, electric charge, and are made of quarks and, thus, experience the strong and weak force.

EVALUATE Understanding how particles interact helps us understand the forces between them.

2: Charge of quarks

Imagine that you are unaware of the charge of the up and down quarks, but you do know that two up quarks and one down quark make up the proton and one up quark and two down quarks make up the neutron. What are the charges of the up and down quarks?

IDENTIFY AND SET UP Protons have charge $+e$ and neutrons are neutral. We'll combine the quarks' charges, set them equal to the charges of the proton and neutron, and then solve for the quarks' charges.

EXECUTE The charges of the quarks in the proton add to $+1e$, giving

$$2q_u + 1q_d = +1e.$$

The charges of the quarks in the neutron add to 0, giving

$$1q_u + 2q_d = 0.$$

We use the neutron result to substitute into the proton result to solve for the charge on the up quarks, yielding

$$2q_u - \tfrac{1}{2}q_u = +1e,$$
$$\tfrac{3}{2}q_u = +1e,$$
$$q_u = +\tfrac{2}{3}e.$$

We then find that the charge on the down quarks is $-e/3$. So both up and down quarks have fractional charges of e!

EVALUATE A process similar to the one we used in this problem led to the discovery that quarks have fractional charge.

Problems

1: Rho meson decay

A neutral rho meson (ρ^0, mass 770 MeV/c^2) decays at rest into a pair of charged pions (π^+, π^-, masses 140 MeV/c^2). Find the energies and momenta of the pions.

IDENTIFY AND SET UP The rest mass of the rho meson will be converted to the rest mass and kinetic energy of the pions. Momentum must be conserved, so each pion carries away equal momentum and energy.

EXECUTE A particle's relativistic energy is

$$E = \sqrt{m^2 c^4 + p^2 c^2}.$$

The rest mass of the rho meson is shared between the two pions. Each pion will have half the total energy of the rho meson:

$$E_{\text{pion}} = \tfrac{1}{2} M_{\text{rho}} c^2 = \tfrac{1}{2}(770 \text{ MeV/c}^2)c^2 = 385 \text{ MeV}.$$

The momentum is found from the relativistic energy:

$$E_{\text{pion}} = \sqrt{m_{\text{pion}}^2 c^4 + p_{\text{pion}}^2 c^2}.$$

Solving for the momentum yields

$$P_{\text{pion}} = \frac{1}{c^2}\sqrt{E_{\text{pion}}^2 - m_{\text{pion}}^2 c^4} = \frac{1}{c}\sqrt{(385 \text{ MeV})^2 - (140 \text{ MeV/c}^2)^2 c^4} = 359 \text{ MeV/c}.$$

Each pion has an energy of 385 MeV and a momentum of 359 MeV/c.

KEY CONCEPT 6 **EVALUATE** This problem illustrates how physicists analyze elementary particle decays and how they must include relativistic kinematics in their work. They also check charge conservation, which holds in this case because the neutral rho meson decays into a positive and negative pair of pions.

2: Z_0 production

A Z_0 particle, one of the mediators of the weak interaction, has a mass of $91 \text{ GeV}/c^2$. It can be produced at rest in a high energy electron–positron collider in which electrons and positrons in counter-rotating beams are allowed to collide. Estimate the product of the magnetic field and beam radius rB necessary to create Z_0 particles.

IDENTIFY AND SET UP We'll calculate the energy needed by the electrons and positrons to create the Z_0 particles. Then we'll find the design parameter rB.

EXECUTE To create Z_0 particles, electrons and positrons are annihilated in a head-on collision. Each particle must carry half the rest energy of the Z_0 particle, so each must have 45.5 GeV of energy. The rest energy of the electron and positron is very small compared with 45.5 GeV, so the momentum is simply

$$p = \frac{E}{c} = 45.5 \text{ GeV/c}.$$

The magnetic field necessary to keep a particle in a magnetic field at a fixed radius is

$$r = \frac{p}{qB}.$$

This expression holds for these relativistic particles. Solving for rB gives

$$rB = \frac{p}{q}$$
$$= \frac{45.5 \text{ GeV/c}}{e}$$
$$= \frac{(45.5 \text{ GeV})(1.6 \times 10^{-19} \text{ J/eV})}{(1.6 \times 10^{-19} \text{ C})(3 \times 10^8 \text{ m/s})} = 152 \text{ m} \cdot \text{T}.$$

KEY CONCEPT 1 **EVALUATE** With 0.1-T magnets, how large a ring is necessary? The ring's radius would need to be 1.5 km.

Practice Problem: Using strong refrigerator magnets $(B = 100 \text{ G})$, what would the radius need to be? *Answer:* 15.2 km

Extra Practice: What magnetic field is required to have a ring radius of 5 m? *Answer:* 30.4 T

3: Possible decays

Consider the following decays of the positively charged pion:

(1) $\pi^+ \to \mu^+ \nu_\mu$

(2) $\pi^+ \to e^+ \nu_e$

(3) $\pi^+ \to \mu^+ \nu_\mu \gamma$

(4) $\pi^+ \to e^+ \nu_e \gamma$

(5) $\pi^+ \to e^+ \nu_e \pi^0$

(6) $\pi^+ \to e^+ \nu_e e^+ e^-$

(7) $\pi^+ \to e^+ \nu_e \nu \bar{\nu}$

(8) $\pi^+ \to \mu^+ \bar{\nu}_e$

(9) $\pi^+ \to \mu^+ \nu_e$

(10) $\pi^+ \to \mu^- e^+ e^- \nu$

Which decays are allowed by conservation of lepton number?

IDENTIFY AND SET UP We'll use Table 2 in the text to determine the lepton numbers of the particles. We'll then check whether the lepton numbers are conserved.

EXECUTE For all decays, the positive pion has no lepton number, so we will check whether there is a net lepton number in the decay products. In decay (1), the muon numbers are −1 and 1, adding to zero. In decay (2), the electron numbers add to zero. In decays (3)–(5), the third particle has no lepton number, so the overall lepton numbers add to zero. In decays (6) and (7), there are four electron numbers, but they add to zero. So the first seven decays conserve lepton number and are allowed.

Decay (8) doesn't conserve lepton number, as the right side has −1 for the positive muon and −1 for the antineutrino. Decays (9) and (10) conserve overall lepton number, but don't conserve electron or muon numbers.

KEY CONCEPT 4 **EVALUATE** Decays (1) through (6) have been observed in particle accelerators. Decays (7) through (10) have not been observed.

4: Size of the universe

According to Hubble's law, matter at a distance r travels away from us at a speed

$$v = H_0 r,$$

where $H_0 = 2.3 \times 10^{-18}$ /s is Hubble's constant. What is the age of the universe according to Hubble's law?

IDENTIFY AND SET UP If we assume that Hubble's constant has remained constant throughout the lifetime of the universe, then the age of the universe is the size of the universe divided by its velocity of expansion. We'll use this method to find the age of the universe.

EXECUTE The age of the universe is

$$T = \frac{r}{v}.$$

Substituting the velocity from Hubble's law, we find that

$$T = \frac{r}{H_0 r} = \frac{1}{H_0} = \frac{1}{2.3 \times 10^{-18} \text{ /s}} = 4.35 \times 10^{17} \text{ s} = 13.8 \text{ billion years.}$$

Hubble's law predicts that the age of the universe is roughly 14 billion years old.

KEY CONCEPT 9 **EVALUATE** This is the currently accepted value for the age of the universe. There is evidence, however, that the universe is expanding at an accelerating rate, meaning that Hubble's constant is not constant. We are finding that the universe is even more interesting than we once thought.

Practice Problem: How fast is a galaxy moving away from us if it is 500,000 light years away? *Answer:* 10.0 km/s

Extra Practice: How far away is a galaxy if it is moving away at 30 km/s? *Answer:* 1.38 million light years

Try It Yourself!

1: Protons in a cyclotron

Protons in a cyclotron spiral out to a radius of 15 cm. The magnetic field has a magnitude of 1.25 T in the cyclotron. (a) Find the frequency of the alternating voltage used to accelerate the protons in the gap. (b) Find the energy of the protons.

Solution Checkpoints

IDENTIFY AND SET UP Use the cyclotron frequency and energy conservation to solve.

EXECUTE (a) The cyclotron frequency, or angular frequency of rotation, is given by

$$\omega = \frac{Be}{m}.$$

This corresponds to a voltage frequency of 1.91×10^7 Hz.
(b) The kinetic energy of the protons is

$$\frac{1}{2} m (\omega r)^2 = 1.6 \text{ MeV.}$$

KEY CONCEPT 1 **EVALUATE** The properties of the magnetic force in a uniform magnetic field. Introductory physics is used by particle physicists every day.

2: Quark flavors

Find the charge and strangeness of all mesons that can be constructed from quark–antiquark pairs with flavors *u*, *d*, and *s*.

Solution Checkpoints

IDENTIFY AND SET UP Find all possible combinations of the three quarks and three antiquarks.

EXECUTE There are nine independent combinations of these quarks, given in the following table (the charge and strangeness are found by adding the charge and strangeness of the quarks that make up the meson):

From Chapter 44 of Student's Study Guide to accompany *University Physics with Modern Physics, Volume 3,* Fifteenth Edition.
Hugh D. Young and Roger A. Freedman. Copyright © 2020 by Pearson Education, Inc. All rights reserved.

Combination	Charge	Strangeness
$u\bar{u}, d\bar{d}, s\bar{s}$	0	0
$u\bar{d}$	+1	0
$d\bar{u}$	−1	0
$u\bar{s}$	+1	+1
$s\bar{u}$	−1	−1
$s\bar{d}$	0	−1
$d\bar{s}$	0	+1

Table 1 Try It Yourself 2.

EVALUATE The second and third particles are the positive and negative pions. The last four particles are called kaons and are four varieties of strange mesons.

Key Example Variation Problems

Solutions to these problems are in Chapter 44 of the Student's Solutions Manual.

Be sure to review EXAMPLE 44.1 (Section 44.2) before attempting these problems.

VP44.1.1 In a cyclotron the maximum radius for the path of protons is 0.800 m. The magnetic field is 0.600 T. Find (a) the frequency of the alternating voltage applied to the dees, (b) the maximum proton kinetic energy in MeV, and (c) the maximum proton speed in terms of c.

VP44.1.2 A cyclotron at the TRIUMF laboratory in Vancouver, Canada, has a magnetic field of 0.460 T. It accelerates negative hydrogen ions with charge $-e$ and mass 1.67×10^{-27} kg. Find (a) the radius of the path of the ions when they have speed 6.00×10^{6} m/s and (b) the angular frequency of the ions' motion.

VP44.1.3 The first cyclotron (in 1930) had a diameter of 11.4 cm and was used to accelerate negative hydrogen ions (charge $-e$, mass 1.67×10^{-27} kg) to kinetic energies of 80.0 keV. Find (a) the magnetic field used and (b) the frequency of the alternating voltage applied to the dees.

VP44.1.4 A cyclotron is used to accelerate alpha particles (charge $+2e$, mass 6.64×10^{-27} kg). The frequency of the alternating voltage applied to the dees is 5.80 MHz. Find (a) the magnetic field used and (b) the kinetic energy, in MeV, of the alpha particles when the radius of their circular path is 0.650 m.

Be sure to review EXAMPLES 44.2 and 44.3 (Section 44.2) before attempting these problems.

VP44.3.1 A proton with kinetic energy K collides with a proton at rest. Both protons remain after the collision along with an eta-prime (η') particle that has rest energy 958

MeV. In order for this reaction to happen, what must be (a) the minimum available energy in the center-of-momentum system and (b) the minimum value of K?

VP44.3.2 A proton with kinetic energy K collides with a proton at rest. The particles that remain after the collision are one proton and a delta-plus (Δ^+) that has rest energy 1232 MeV. In order for this reaction to happen, what must be (a) the minimum available energy in the center-of-momentum system and (b) the minimum value of K?

VP44.3.3 A negative pion (π^-) that has rest energy 140 MeV and kinetic energy K collides with a proton at rest. What remains after the collision is a delta-zero (Δ^0) that has rest energy 1232 MeV. In order for this reaction to happen, what must be (a) the minimum available energy in the center-of-momentum system and (b) the minimum value of K?

VP44.3.4 The Large Hadron Collider gives each proton a total energy of $7 \text{ TeV} = 7.00 \times 10^3$ GeV. (a) If a proton that has this energy collides with a second proton that is stationary, what is the available energy? (b) After the collision, both protons remain as well as a neutral particle X. What is the maximum mass of X?

Be sure to review EXAMPLES 44.8 and 44.9 (Section 44.6) before attempting these problems.

VP44.9.1 The red spectral line of hydrogen has a wavelength of 656.3 nm when the hydrogen is at rest. This line is observed to have a wavelength of 725 nm in the spectrum of a galaxy. Find (a) the redshift z and (b) the recession speed of the galaxy in terms of c.

VP44.9.2 A prominent absorption line in the spectrum of the sun (see Fig. 39.9) and many galaxies is due to singly ionized calcium. In the rest frame, this line has an ultraviolet wavelength of 396.9 nm. If a galaxy is receding from us at $0.0711c$, find (a) the wavelength at which we'll observe this line and (b) the redshift z.

VP44.9.3 Find the distances to (a) a galaxy with a recession speed of $0.0992c$ and (b) a galaxy with a recession speed of $0.0711c$. Give your answers in pc.

VP44.9.4 The spectrum of a Type Ia supernova (an exploding star so luminous that it can be seen even in distant galaxies) has an emission line due to singly ionized silicon. If astronomers observe this emission line at 615.0 nm in the laboratory but 666 nm in the spectrum of a Type Ia supernova in a distant galaxy, find (a) the recession speed of the galaxy in m/s and (b) the distance to the galaxy in pc.

From Chapter 44 of Student's Study Guide to accompany *University Physics with Modern Physics, Volume 3,* Fifteenth Edition.
Hugh D. Young and Roger A. Freedman. Copyright © 2020 by Pearson Education, Inc. All rights reserved.

STUDENT'S SOLUTIONS MANUAL FOR PARTICLE PHYSICS AND COSMOLOGY

VP44.1.1. **IDENTIFY:** This problem is about a cyclotron accelerating protons.
SET UP and EXECUTE: (a) We want the frequency. Use Eq. (44.7).

$$f = \frac{eB}{2\pi m} = \frac{e(0.600\ \text{T})}{2\pi m_\text{p}} = 9.15\ \text{MHz}.$$

(b) We want the maximum kinetic energy of the proton. Use Eq. (44.8) with $B = 0.600$ T and $R = 0.800$ m.

$$K_\text{max} = \frac{e^2 B^2 R^2}{2m_\text{p}} = 11.0\ \text{MeV}.$$

(c) We want the proton's speed. The rest energy of the proton is 938 MeV which is much greater than the kinetic energy of 11 MeV, so we do not have to use the relativistic equation. Solving $K = \frac{1}{2}mv^2$ for v gives

$$v = \sqrt{2K/m_\text{p}} = 4.60 \times 10^7\ \text{m/s} = 0.153c.$$

EVALUATE: Since $v \approx 15\%$ of c, it is reasonable not to use relativity. If we had used relativity, our result would be

$$K = mc^2(\gamma - 1) \rightarrow \gamma = 1.012$$
$$v = c\sqrt{1 - 1/\gamma^2} = 0.152c.$$

The percent difference between the two answers is $0.001/0.152 = 0.66\%$, which is extremely small.

VP44.1.2. **IDENTIFY:** This problem is about a cyclotron accelerating hydrogen ions (protons).
SET UP and EXECUTE: (a) We want the radius R. $R = m_\text{p}v/eB$. Using $B = 0.460$ T and the given speed, we get $R = 0.136$ m.
(b) We want the angular frequency. Use Eq. (44.7) with $q = e$ and $B = 0.460$ T, which gives $\omega = eB/m = 4.41 \times 10^7$ rad/s.
EVALUATE: This angular frequency is rather small compared to many modern cyclotrons.

VP44.1.3. **IDENTIFY:** This problem deals with the operation of a cyclotron.
SET UP and EXECUTE: (a) We want the magnetic field. The rest energy of the proton is 938 MeV and its kinetic energy here is 80.0 keV, which is much less than its rest energy. Therefore we do not need to use special relativity. We use Eq. (44.8), with $R = (11.4\ \text{cm})/2$ and $K_\text{max} = 80.0$ keV, and solve for B, which gives

$$B = \frac{\sqrt{2m_\text{p}K_\text{max}}}{eR} = 0.717\ \text{T}.$$

(b) We want the frequency. Use Eq. (44.7) with $B = 0.717$ T and $q = e$.

$$f = \frac{eB}{2\pi m} = \frac{e(0.717 \text{ T})}{2\pi m_p} = 10.9 \text{ MHz}.$$

EVALUATE: Modern cyclotrons produce much greater kinetic energy, but after all, this was the first one.

VP44.1.4. **IDENTIFY:** This problem deals with the operation of a cyclotron.
SET UP and **EXECUTE:** **(a)** We want the magnetic field and know that the frequency is 5.80 MHz. Use Eq. (44.7) and solve for B using $q = 2e$ and the given mass of an alpha particle.

$$B = \frac{2\pi m_\alpha f}{2e} = 0.756 \text{ T}.$$

(b) We want the kinetic energy when $R = 0.650$ m. Using Eq. (44.8) with $B = 0.756$ T gives

$$K_{max} = \frac{(2e)^2 B^2 R^2}{2m_\alpha} = 11.6 \text{ MeV}.$$

EVALUATE: A magnetic field of 0.756 T is certainly physically reasonable in a physics laboratory.

VP44.3.1. **IDENTIFY:** This problem deals with the available energy when particles collide.
SET UP and **EXECUTE:** We follow the procedure of Example 44.2.
(a) We want the available energy E_a. The total available energy must be the total rest energy in the center-of-momentum frame.

$$E_a = 2m_p c^2 + m_{n'} c^2 = 2(938 \text{ MeV}) + 958 \text{ MeV} = 2834 \text{ MeV}.$$

(b) We want the kinetic energy K of the incoming proton. First use Eq. (44.10) to find E_m (the total energy of the incoming proton). Then use $E_m = K + m_p c^2$ to find K.

$$E_m = \frac{E_a^2}{2m_p c^2} - m_p c^2 = K + m_p c^2$$

$$K = \frac{E_a^2}{2m_p c^2} - 2m_p c^2$$

Using $E_a = 2834$ MeV and $m_p c^2 = 938$ MeV gives us $K = 2410$ MeV $= 2.41$ GeV.
EVALUATE: Colliding beams would require far less kinetic energy because nearly all the incoming kinetic energy is available energy.

VP44.3.2. **IDENTIFY:** This problem deals with the available energy during collisions.
SET UP and **EXECUTE:** **(a)** We want E_a. The minimum available energy is the rest energy of the particles after the collision. $E_a = m_p c^2 + m_\Delta c^2 = 938$ MeV + 1232 MeV = 2170 MeV = 2.17 GeV.
(b) We want the kinetic energy. Use the same method as in problem VP44.3.1.

$$K = \frac{E_a^2}{2m_p c^2} - 2m_p c^2$$

Using $E_a = 2170$ MeV and $m_p c^2 = 938$ MeV gives $K = 634$ MeV.
EVALUATE: The incoming proton could have more than 634 MeV of kinetic energy. In that case, the products would each have kinetic energy.

VP44.3.3. **IDENTIFY:** This problem deals with the available energy during collisions.
SET UP and **EXECUTE:** **(a)** We want E_a. The minimum available energy is the rest energy of the product (the Δ^0), which is 1232 MeV.
(b) We want the minimum kinetic energy K of the pion. The target and incident particle have different masses, so use Eq. (44.9) to find the available energy. Then use this to find E_m, and finally use $E_m = K + mc^2$ to find K.
The quantities in Eq. (44.9) are:
$$M = \text{target} = \text{proton, so } Mc^2 = 938 \text{ MeV}$$

$$m = \text{incident particle} = \pi^-, \text{so } mc^2 = 140 \text{ MeV}$$

Using the result of part (a), Eq. (44.9) gives

$$(1232 \text{ MeV})^2 = 2(938 \text{ MeV})E_m + (938 \text{ MeV})^2 + (140 \text{ MeV})^2$$
$$E_m = 329.6 \text{ MeV}.$$

Now use $E_m = K + mc^2$ to find K.

$$329.6 \text{ MeV} = K + 140 \text{ MeV}, \text{ which gives } K = 190 \text{ MeV}.$$

EVALUATE: Notice that the kinetic energy is a significant fraction of E_m.

VP44.3.4. IDENTIFY: This problem deals with the available energy during collisions.
SET UP and EXECUTE: (a) We want E_a. $E_m = 7000$ GeV and $m_p c^2 = 938$ MeV, so $E_m \gg m_p c^2$. Therefore we use Eq. (44.11). Using the given rest energies gives

$$E_a = \sqrt{2\,mc^2 E_m} = \sqrt{2(938 \text{ MeV})(7000 \text{ GeV})} = 115 \text{ GeV}.$$

(b) We want the minimum mass of particle X. E_a is at *least* equal to the rest energy of the products of the collision, which are 2 protons and particle X. So 115 GeV = 2(938 MeV) + $m_X c^2$. This gives $m_X = 113$ GeV/c^2.
EVALUATE: Particle X is 120 times heavier than the proton.

VP44.9.1. IDENTIFY: This problem concerns the red shift.
SET UP and EXECUTE: (a) The target variable is z.

$$z = \frac{\Delta\lambda}{\lambda} = \frac{725 \text{ nm} - 656.3 \text{ nm}}{656.3 \text{ nm}} = 0.105.$$

(b) We want the speed. Use Eq. (44.14) with the given wavelengths.

$$v = \frac{(\lambda_0/\lambda_s)^2 - 1}{(\lambda_0/\lambda_s)^2 + 1}c = \frac{(725/656.3)^2 - 1}{(725/656.3)^2 + 1}c = 0.0992c.$$

EVALUATE: The fact that we have a red shift instead of a blue shift tells us that the galaxy is receding from us at 9.92% the speed of light.

VP44.9.2. IDENTIFY: This problem is about the red shift of a galaxy.
SET UP: Eq. (44.13) applies.
EXECUTE: (a) We want wavelength that we observe. Use Eq. (44.13).

$$\lambda_0 = \lambda_s \sqrt{\frac{c+v}{c-v}} = (396.9 \text{ nm})\sqrt{\frac{c + 0.0711c}{c - 0.0711c}} = 426 \text{ nm}.$$

(b) We want z.

$$z = \frac{\Delta\lambda}{\lambda} = \frac{426 \text{ nm} - 396.9 \text{ nm}}{396.9 \text{ nm}} = 0.0738.$$

EVALUATE: Since v is only about 7% the speed of light, the red shift is not very large.

VP44.9.3. IDENTIFY: This problem requires the use of Hubble's law.
SET UP: Hubble's law: $v = H_0 r$, where $H_0 = (68 \text{ km/s})/\text{Mpc}$. The target variable is the distance to the galaxy.
EXECUTE: (a) $v = 0.0992c$. Apply Hubble's law using the measured speed.

$$r = \frac{v}{H_0} = \frac{0.0992c}{68 \dfrac{\text{km/s}}{\text{Mpc}}} = 440 \text{ Mpc} = 4.4 \times 10^8 \text{ pc.}$$

(b) $v = 0.0711c$. Use the same approach as in part (a).

$$r = \frac{v}{H_0} = \frac{0.0711c}{68 \dfrac{\text{km/s}}{\text{Mpc}}} = 310 \text{ Mpc} = 3.1 \times 10^8 \text{ pc.}$$

EVALUATE: Speeds can be determined quite easily using spectral analysis, but H_0 is not so easy to measure. These distances are only as accurate as H_0.

VP44.9.4. **IDENTIFY:** This problem requires the use of Hubble's law and the red shift.

SET UP: Hubble's law: $v = H_0 r$, where $H_0 = (68 \text{ km/s})/\text{Mpc}$.

EXECUTE: (a) We want the speed of recession of the galaxy. Use Eq. (44.14).

$$v = \frac{(\lambda_0/\lambda_s)^2 - 1}{(\lambda_0/\lambda_s)^2 + 1} c = \frac{(666/615)^2 - 1}{(666/615)^2 + 1} c = 0.0795c = 2.38 \times 10^7 \text{ m/s}.$$

(b) We want the distance to the galaxy. Use Hubble's law.

$$r = \frac{v}{H_0} = \frac{2.38 \times 10^7 \text{ m/s}}{68 \dfrac{\text{km/s}}{\text{Mpc}}} = 350 \text{ Mpc} = 3.5 \times 10^8 \text{ pc}.$$

EVALUATE: Using 1 pc = 3.26 ly, we find that this galaxy is 1.1 billion light-years from us!

44.3. **IDENTIFY:** The energy released is the energy equivalent of the mass decrease that occurs in the decay.

SET UP: The mass of the pion is $m_{\pi^+} = 270m_e$ and the mass of the muon is $m_{\mu^+} = 207m_e$. The rest energy of an electron is 0.511 MeV.

EXECUTE: (a) $\Delta m = m_{\pi+} - m_{\mu+} = 270m_e - 207m_e = 63m_e \Rightarrow E = 63(0.511 \text{ MeV}) = 32 \text{ MeV}.$

EVALUATE: (b) A positive muon has less mass than a positive pion, so if the decay from muon to pion was to happen, you could always find a frame where energy was not conserved. This cannot occur.

44.5. **IDENTIFY:** The kinetic energy of the alpha particle is due to the mass decrease.

SET UP and EXECUTE: $^1_0 n + {}^{10}_5 B \rightarrow {}^7_3 Li + {}^4_2 He$. The mass decrease in the reaction is

$m(^1_0 n) + m(^{10}_5 B) - m(^7_3 Li) - m(^4_2 He) = 1.008665 \text{ u} + 10.012937 \text{ u} - 7.016005 \text{ u} - 4.002603 \text{ u}$
$= 0.002994 \text{ u}$ and the energy released is $E = (0.002994 \text{ u})(931.5 \text{ MeV/u}) = 2.79 \text{ MeV}$. Assuming

the initial momentum is zero, $m_{Li} v_{Li} = m_{He} v_{He}$ and $v_{Li} = \dfrac{m_{He}}{m_{Li}} v_{He}$. $\dfrac{1}{2} m_{Li} v_{Li}^2 + \dfrac{1}{2} m_{He} v_{He}^2 = E$

becomes

$$\frac{1}{2} m_{Li} \left(\frac{m_{He}}{m_{Li}} \right)^2 v_{He}^2 + \frac{1}{2} m_{He} v_{He}^2 = E \text{ and } v_{He} = \sqrt{\frac{2E}{m_{He}} \left(\frac{m_{Li}}{m_{Li} + m_{He}} \right)}. \quad E = 4.470 \times 10^{-13} \text{ J}.$$

$$m_{He} = 4.002603 \text{ u} - 2(0.0005486 \text{ u}) = 4.0015 \text{ u} = 6.645 \times 10^{-27} \text{ kg}.$$

$m_{Li} = 7.016005 \text{ u} - 3(0.0005486 \text{ u}) = 7.0144 \text{ u}$. This gives $v_{He} = 9.26 \times 10^6 \text{ m/s}.$

EVALUATE: The speed of the alpha particle is considerably less than the speed of light, so it is not necessary to use the more complicated relativistic formulas.

44.7. **IDENTIFY:** This problem is about the available energy during a collision of equal-mass particles.

SET UP: We are comparing the available kinetic energy if the target is stationary or if we have colliding beams.

EXECUTE: (a) Stationary target. Use Eq. 44.10 and solve for K.

$$E_a^2 = 2mc^2(mc^2 + K)$$

$$K = \frac{E_a^2}{2mc^2} - mc^2 = \frac{(2 \text{ TeV})^2}{2(938 \text{ MeV})} - 938 \text{ MeV} = 2130 \text{ TeV}.$$

(b) Colliding beams. The available energy is the kinetic energy minus the rest energy of the two protons.

$$E_a = 2K - 2m_p c^2$$

$$K = \frac{E_a + 2m_p c^2}{2} = \frac{2.00 \text{ TeV} + 2(938 \text{ MeV})}{2} = 1.00 \text{ TeV}.$$

EVALUATE: By using colliding beams we need less than 1/1000 the kinetic energy than with a stationary target.

44.11. IDENTIFY and SET UP: The masses of the target and projectile particles are equal, so we can use the equation $E_a^2 = 2mc^2(E_m + mc^2)$. E_a is specified; solve for the energy E_m of the beam particles.

EXECUTE: **(a)** Solve for E_m: $E_m = \frac{E_a^2}{2mc^2} - mc^2.$

The mass for the alpha particle can be calculated by subtracting two electron masses from the $_2^4$He atomic mass:

$$m = m_\alpha = 4.002603 \text{ u} - 2(0.0005486 \text{ u}) = 4.001506 \text{ u}.$$

Then $mc^2 = (4.001506 \text{ u})(931.5 \text{ MeV/u}) = 3.727 \text{ GeV}.$

$$E_m = \frac{E_a^2}{2mc^2} - mc^2 = \frac{(16.0 \text{ GeV})^2}{2(3.727 \text{ GeV})} - 3.727 \text{ GeV} = 30.6 \text{ GeV}.$$

(b) Each beam must have $\frac{1}{2}E_a = 8.0 \text{ GeV}.$

EVALUATE: For a stationary target the beam energy is nearly twice the available energy. In a colliding beam experiment all the energy is available and each beam needs to have just half the required available energy.

44.15. IDENTIFY and SET UP: For the reaction $p + p \rightarrow p + p + p + \bar{p}$, the two incident protons must have enough kinetic energy to produce a p and a $\bar{p}$, plus any kinetic energy of the products. If they have the minimum kinetic energy, the products are at rest. The proton and antiproton have equal masses. The available energy for two equal-mass particles is $E_a^2 = 2mc^2(E_m + mc^2)$, where $E_m = K + mc^2$.

EXECUTE: **(a)** In a head-on collision with equal speeds, the laboratory frame is the center-of-momentum frame. For the minimum kinetic energy of the incident protons, the products are all at rest. In that case, the incident protons need only enough kinetic energy to produce a proton and an antiproton. Since the incident protons have equal energy, each one must have kinetic energy equal to the rest energy of a proton, which is 938 MeV.

(b) In this case, the target proton is at rest. Since 4 particles are produced, each of mass m, the available energy E_a must be at least equal to $4mc^2$. Therefore $E_a^2 = 2mc^2(E_m + mc^2) = (4mc^2)^2$ $= 16m^2c^4$, which gives $E_m = 7mc^2$. Using $E_m = K + mc^2$, we get $K = 6mc^2 = 6(938 \text{ MeV}) = 5630 \text{ MeV}.$

EVALUATE: When the two protons collide head-on with equal speeds, they need only 938 MeV of kinetic energy each, for a total of 1879 MeV. But when the target is stationary, the kinetic energy needed is 5630 MeV, which is 3 times as much as for a head-on collision.

44.17. IDENTIFY: The kinetic energy comes from the mass decrease.

SET UP: Table 44.3 gives $m(K^+) = 493.7 \text{ MeV}/c^2$, $m(\pi^0) = 135.0 \text{ MeV}/c^2$, and $m(\pi^\pm)$ $= 139.6 \text{ MeV}/c^2$.

EXECUTE: **(a)** Charge must be conserved, so $K^+ \rightarrow \pi^0 + \pi^+$ is the only possible decay.

(b) The mass decrease is

$m(K^+) - m(\pi^0) - m(\pi^+) = 493.7 \text{ MeV}/c^2 - 135.0 \text{ MeV}/c^2 - 139.6 \text{ MeV}/c^2 = 219.1 \text{ MeV}/c^2$. The energy released is 219.1 MeV.

EVALUATE: The π mesons do not share this energy equally since they do not have equal masses.

44.21. **IDENTIFY** and **SET UP:** Find the energy equivalent of the mass decrease.

EXECUTE: The mass decrease is $m(\Sigma^+) - m(\mathrm{p}) - m(\pi^0)$ and the energy released is

$mc^2(\Sigma^+) - mc^2(\mathrm{p}) - mc^2(\pi^0) = 1189\ \mathrm{MeV} - 938.3\ \mathrm{MeV} - 135.0\ \mathrm{MeV} = 116\ \mathrm{MeV}$. (The mc^2 values for each particle were taken from Table 44.3.)

EVALUATE: The mass of the decay products is less than the mass of the original particle, so the decay is energetically allowed and energy is released.

44.27. **IDENTIFY:** The charge, baryon number, and strangeness of the particles are the sums of these values for their constituent quarks.

SET UP: The properties of the six quarks are given in Table 44.5.

EXECUTE: **(a)** $S = 1$ indicates the presence of one $\bar{s}$ antiquark and no s quark. To have baryon number 0 there can be only one other quark, and to have net charge $+e$ that quark must be a u, and the quark content is $u\bar{s}$.

(b) The particle has an $\bar{s}$ antiquark, and for a baryon number of -1 the particle must consist of three antiquarks. For a net charge of $-e$, the quark content must be $\overline{dd}\,\bar{s}$.

(c) $S = -2$ means that there are two s quarks, and for baryon number 1 there must be one more quark. For a charge of 0 the third quark must be a u quark and the quark content is uss.

EVALUATE: The particles with baryon number zero are mesons and consist of a quark-antiquark pair. Particles with baryon number 1 consist of three quarks and are baryons. Particles with baryon number -1 consist of three antiquarks and are antibaryons.

44.31. **(a) IDENTIFY** and **SET UP:** Hubble's law is $v = H_0 r$, with $H_0 = (67.3\ \mathrm{km/s})/(\mathrm{Mpc})$. 1 Mpc = 3.26 Mly.

EXECUTE: $r = 5210$ Mly, so

$v = H_0 r = [(67.3\ \mathrm{km/s})/\mathrm{Mpc}](1\ \mathrm{Mpc}/3.26\ \mathrm{Mly})(5210\ \mathrm{Mly}) = 1.08 \times 10^5\ \mathrm{km/s} = 1.08 \times 10^8\ \mathrm{m/s}$.

(b) IDENTIFY and **SET UP:** Use v from part (a) in $\lambda_0 = \lambda_S \sqrt{\dfrac{c+v}{c-v}} = \sqrt{\dfrac{1+v/c}{1-v/c}}$.

EXECUTE: $\dfrac{\lambda_0}{\lambda_S} = \sqrt{\dfrac{c+v}{c-v}} = \sqrt{\dfrac{1+v/c}{1-v/c}}$.

$\dfrac{v}{c} = \dfrac{1.08 \times 10^8\ \mathrm{m/s}}{2.998 \times 10^8\ \mathrm{m/s}} = 0.3602$, so $\dfrac{\lambda_0}{\lambda_S} = \sqrt{\dfrac{1+0.3602}{1-0.3602}} = 1.46$.

EVALUATE: The galaxy in Examples 44.8 and 44.9 is 710 Mly away so has a smaller recession speed and redshift than the galaxy in this problem.

44.35. **IDENTIFY:** The reaction energy Q is defined in Chapter 43 as $Q = (M_A + M_B - M_C - M_D)c^2$ and is the energy equivalent of the mass change in the reaction. When Q is negative the reaction is endoergic. When Q is positive the reaction is exoergic.

SET UP: 1 u is equivalent to 931.5 MeV. Use the neutral atom masses that are given in Table 43.2.

EXECUTE: $m_{^{12}_{6}\mathrm{C}} + m_{^{4}_{2}\mathrm{He}} - m_{^{16}_{8}\mathrm{O}} = 7.69 \times 10^{-3}\,\mathrm{u}$, or 7.16 MeV, an exoergic reaction.

EVALUATE: 7.16 MeV of energy is released in the reaction.

44.37. **IDENTIFY:** The energy comes from a mass decrease.

SET UP: A charged pion decays into a muon plus a neutrino. The muon in turn decays into an electron or positron plus two neutrinos.

EXECUTE: **(a)** $\pi^- \to \mu^- + \mathrm{neutrino} \to \mathrm{e}^- + \mathrm{three\ neutrinos}$.

(b) If we neglect the mass of the neutrinos, the mass decrease is

$$m(\pi^-) - m(\mathrm{e}^-) = 273 m_{\mathrm{e}} - m_{\mathrm{e}} = 272 m_{\mathrm{e}} = 2.480 \times 10^{-28}\ \mathrm{kg}.$$

$$E = mc^2 = 2.23 \times 10^{-11} \text{ J} = 139 \text{ MeV}.$$

(c) The total energy delivered to the tissue is $(50.0 \text{ J/kg})(10.0 \times 10^{-3} \text{ kg}) = 0.500 \text{ J}$. The number of π^- mesons required is $\dfrac{0.500 \text{ J}}{2.23 \times 10^{-11} \text{ J}} = 2.24 \times 10^{10}$.

(d) The RBE for the electrons that are produced is 1.0, so the equivalent dose is
$$1.0(50.0 \text{ Gy}) = 50.0 \text{ Sv} = 5.0 \times 10^3 \text{ rem}.$$

EVALUATE: The π are heavier than electrons and therefore behave differently as they hit the tissue.

44.39. **IDENTIFY:** We are investigating colliding proton beams in the Large Hadron Collider.
SET UP and EXECUTE: **(a)** With one bunch per 25 ns, the number of bunches per second is $1/(25 \text{ ns}) = 40$ million.
(b) The fraction of protons that collide is $20/(115 \text{ billion}) = 1.7 \times 10^{-10}$.
(c) We want the number of collisions that occur each second. Using the given information and the result of part (a) gives (40 million bunches/s)(20 collisions/bunch) = 800 million collisions.
(d) We want the proton density ρ in a bunch, with N protons in a cylinder of length L and radius R.
$$\rho = \frac{N}{\pi R^2 L} = \frac{115 \text{ billion}}{\pi (10 \text{ } \mu\text{m})^2 (0.300 \text{ m})} = 1.2 \times 10^{12} \text{ protons/mm}^3.$$
(e) We want the density of hadrons in ordinary matter. Follow the hint. Mass is 75 kg. The body is a cylinder 30 cm in diameter of length 1.75 m tall. $V = \pi R^2 L = \pi (0.15 \text{ m})^2 (1.75 \text{ m}) = 0.124 \text{ m}^3$. The number N of hadrons is $N = (75 \text{ kg})/(1.67 \times 10^{-27} \text{ kg}) = 4.5 \times 10^{28}$ hadrons. The density is $\rho = N/V = (4.5 \times 10^{28})/(0.124 \text{ m}^3) = 3.6 \times 10^{29}$ hadrons/m$^3 \approx 4 \times 10^{20}$ hadrons/mm^3.
EVALUATE: As a check for part (e) we can use the density of water, which is 1000 kg/m^3. This gives
$$\frac{1000 \text{ kg/m}^3}{1.67 \times 10^{-27} \text{ kg}} \approx 6 \times 10^{29} \text{ hadrons/m}^3 = 6 \times 10^{20} \text{ hadrons/mm}^3.$$
This result is quite close to our calculation above. Note that the concentration of hadrons in ordinary matter is around 300 million times as great as in the bunches in the Large Hadron Collider.

44.43. **IDENTIFY and SET UP:** Apply the Heisenberg uncertainty principle in the form $\Delta E \Delta t \approx \hbar/2$. Let ΔE be the energy width and let Δt be the lifetime.

EXECUTE: $\dfrac{\hbar}{2\Delta E} = \dfrac{(1.054 \times 10^{-34} \text{ J} \cdot \text{s})}{2(4.4 \times 10^6 \text{ eV})(1.6 \times 10^{-19} \text{ J/eV})} = 7.5 \times 10^{-23} \text{ s}.$

EVALUATE: The shorter the lifetime, the greater the energy width.

44.45. **IDENTIFY:** Apply $\left|\dfrac{dN}{dt}\right| = \lambda N$ to find the number of decays in one year.

SET UP: Water has a molecular mass of 18.0×10^{-3} kg/mol.
EXECUTE: **(a)** The number of protons in a kilogram is
$$(1.00 \text{ kg})\left(\frac{6.022 \times 10^{23} \text{ molecules/mol}}{18.0 \times 10^{-3} \text{ kg/mol}}\right)(2 \text{ protons/molecule}) = 6.7 \times 10^{25}.$$ Note that only the protons in the hydrogen atoms are considered as possible sources of proton decay. The energy per decay is $m_p c^2 = 938.3 \text{ MeV} = 1.503 \times 10^{-10}$ J, and so the energy deposited in a year, per kilogram, is
$$(6.7 \times 10^{25})\left(\frac{\ln 2}{1.0 \times 10^{18} \text{ y}}\right)(1 \text{ y})(1.50 \times 10^{-10} \text{ J}) = 7.0 \times 10^{-3} \text{ Gy} = 0.70 \text{ rad}.$$

(b) For an RBE of unity, the equivalent dose is $(1)(0.70 \text{ rad}) = 0.70$ rem.

EVALUATE: The equivalent dose is much larger than that due to the natural background. It is not feasible for the proton lifetime to be as short as 1.0×10^{18} y.

44.47. **IDENTIFY:** We are dealing with muons in cosmic rays. The energies involved are much greater than the rest energy of muons, so we must use the relativistic equations.

SET UP and EXECUTE: **(a)** We want the muon's speed. First find γ and then use it to find the speed.

$$E = m\gamma c^2 \quad \rightarrow \quad 6.000 \text{ GeV} = \gamma\,(105.7 \text{ MeV}) \quad \rightarrow \quad \gamma = 56.76443$$

$$v = c\sqrt{1 - 1/\gamma^2} = c\sqrt{1 - 1/(56.76443)^2} = 0.99969c = 2.997 \times 10^8 \text{ m/s}.$$

(b) We want the distance the muon travels in one lifetime. $x = vt = (2.997 \times 10^8 \text{ m/s})(2.197 \text{ } \mu s)$ = 658.4 m.

(c) We want the distance to the earth's surface. To the muon, the 15.00 km is Lorentz contracted, so the distance it sees itself traveling is $L = L_0/\gamma = (15.00 \text{ km})/(56.76443) = 0.264$ km = 264 m.

(d) We want the time in the earth's frame. The muon's frame is the proper frame.

$$\Delta t = \gamma \Delta t_0 = (56.76443)(2.197 \text{ } \mu s) = 124.7 \text{ } \mu s.$$

(e) We want the distance traveled as seen from the earth frame. $x = vt = (2.997 \times 10^8 \text{ m/s})(124.7 \text{ } \mu s)$ = 37.4 km.

(f) As observed in the earth frame, the muon is created 15.00 km above the surface and travels 37.4 km, so it survives its trip through 15.00 km of atmosphere and travels downward into the surface of the earth. At such high energy, the muon can penetrate deeply into the surface without hitting anything, so the depth it reaches is 37.4 km − 15.00 km = 22.4 km.

EVALUATE: Without relativistic effects we would have far fewer cosmic rays striking the earth's surface.

44.49. **IDENTIFY:** The kinetic energy comes from the mass difference.

SET UP and EXECUTE: $K_\Sigma = 180$ MeV. $m_\Sigma c^2 = 1197$ MeV. $m_n c^2 = 939.6$ MeV.

$m_\pi c^2 = 139.6$ MeV. $E_\Sigma = K_\Sigma + m_\Sigma c^2 = 180$ MeV $+ 1197$ MeV $= 1377$ MeV. Conservation of the x-component of momentum gives $p_\Sigma = p_{nx}$. Then

$p_{nx}^2 c^2 = p_\Sigma^2 c^2 = E_\Sigma^2 - (m_\Sigma c)^2 = (1377 \text{ MeV})^2 - (1197 \text{ MeV})^2 = 4.633 \times 10^5 \text{ (MeV)}^2$. Conservation of energy gives $E_\Sigma = E_\pi + E_n$. $E_\Sigma = \sqrt{m_n^2 c^4 + p_n^2 c^2} + \sqrt{m_\pi^2 c^4 + p_\pi^2 c^2}$.

$E_\Sigma - \sqrt{m_n^2 c^4 + p_n^2 c^2} = \sqrt{m_\pi^2 c^4 + p_\pi^2 c^2}$. Square both sides:

$E_\Sigma^2 + m_n^2 c^4 + p_{nx}^2 c^2 + p_{ny}^2 c^2 - 2E_\Sigma E_n = m_\pi^2 c^4 + p_\pi^2 c^2$. $p_\pi = p_{ny}$, so

$$E_\Sigma^2 + m_n^2 c^4 + p_{nx}^2 c^2 - 2E_\Sigma E_n = m_\pi^2 c^4 \quad \text{and} \quad E_n = \frac{E_\Sigma^2 + m_n^2 c^4 - m_\pi^2 c^4 + p_{nx}^2 c^2}{2E_\Sigma}.$$

$$E_n = \frac{(1377 \text{ MeV})^2 + (939.6 \text{ MeV})^2 - (139.6 \text{ MeV})^2 + 4.633 \times 10^5 \text{ (MeV)}^2}{2(1377 \text{ MeV})} = 1170 \text{ MeV}.$$

$$K_n = E_n - m_n c^2 = 1170 \text{ MeV} - 939.6 \text{ MeV} = 230 \text{ MeV}.$$

$$E_\pi = E_\Sigma - E_n = 1377 \text{ MeV} - 1170 \text{ MeV} = 207 \text{ MeV}.$$

$$K_\pi = E_\pi - m_\pi c^2 = 207 \text{ MeV} - 139.6 \text{ MeV} = 67 \text{ MeV}.$$

$$p_n^2 c^2 = E_n^2 - m_n^2 c^2 = (1170 \text{ MeV})^2 - (939.6 \text{ MeV})^2 = 4.861 \times 10^5 \text{ (MeV)}^2.$$

The angle θ the velocity of the neutron makes with the $+x$-axis is given by

$$\cos\theta = \frac{p_{nx}}{p_n} = \sqrt{\frac{4.633\times10^5}{4.861\times10^5}} \text{ and } \theta = 12.5° \text{ below the } +x\text{-axis.}$$

EVALUATE: The decay particles do not have equal energy because they have different masses.

44.51. **IDENTIFY and SET UP:** For nonrelativistic motion, the maximum kinetic energy in a cyclotron is $K_{max} = \dfrac{q^2 R^2}{2m} B^2$. The angular frequency is $\omega = |q|B/m$.

EXECUTE: **(a)** The rest energy of a proton is 938 MeV, and the kinetic energies in the data table in the problem are around 1 MeV or less, so there is no need to use relativistic expressions.
(b) Figure 44.51 shows the graph of K_{max} versus B^2 for the data in the problem. The graph is clearly a straight line and has slope equal to 6.748 MeV/T^2 = 1.081 ×10^{-12} J/T^2. The formula for K_{max} is $K_{max} = \dfrac{q^2 R^2}{2m} B^2$, so a graph of K_{max} versus B^2 should be a straight line with slope equal to $q^2 R^2/2m$. Solving $q^2 R^2/2m =$ slope for R gives

$$R = \sqrt{\frac{2m(\text{slope})}{q^2}} = \sqrt{\frac{2(1.673\times10^{-27} \text{ kg})(1.081\times10^{-12} \text{ J/T}^2)}{(1.602\times10^{-19} \text{ C})^2}} = 0.375 \text{ m} = 37.5 \text{ cm.}$$

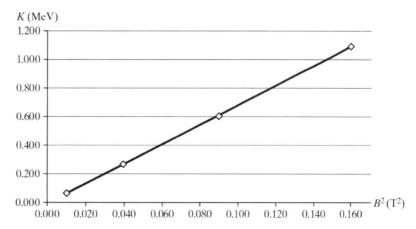

K (MeV)

Figure 44.51

(c) Using the result from our graph, we get $K_{max} = (\text{slope})B^2 = (6.748 \text{ MeV/T}^2)(0.25 \text{ T})^2 = 0.42$ MeV.
(d) The angular speed is $\omega = |q|B/m = (1.602\times10^{-19} \text{ C})(0.40 \text{ T})/(1.67\times10^{-27} \text{ kg}) = 3.8\times10^7$ rad/s.
EVALUATE: In part (c) we can check by using $K_{max} = q^2 R^2 B^2/2m = (qRB)^2/2m$. Using $B = 0.25$ T and the standard values for the other quantities gives $K_{max} = 6.75\times10^{-14}$ J = 0.42 MeV, which agrees with our result.

44.53. **IDENTIFY and SET UP:** Construct the diagram as specified in the problem. In part (b), use quark charges $u = +\dfrac{2}{3}, d = \dfrac{-1}{3}$, and $s = \dfrac{-1}{3}$ as a guide.

EXECUTE: **(a)** The diagram is given in Figure 44.53. The Ω^- particle has $Q = -1$ (as its label suggests) and $S = -3$. Its appears as a "hole" in an otherwise regular lattice in the $S - Q$ plane.
(b) The quark composition of each particle is shown in the figure.

EVALUATE: The mass difference between each S row is around 145 MeV (or so). This puts the Ω^- mass at about the right spot. As it turns out, all the other particles on this lattice had been discovered already and it was this "hole" and mass regularity that led to an accurate prediction of the properties of the Ω^-!

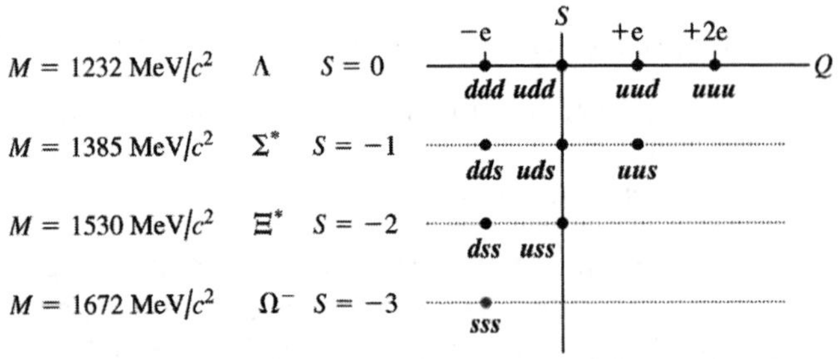

Figure 44.53

44.55. IDENTIFY: We are dealing with an expanding curved space.
SET UP: Follow the directions with each part.
EXECUTE: **(a)** We want D. If θ is in radians, then $D = R\theta$.
(b) We want V. Use $V = dD/dt$ and the result from part (a).

$$V = \frac{dD}{dt} = \frac{d(R\theta)}{dt} = \theta\frac{dR}{dt} = \frac{D}{R}\frac{dR}{dt}.$$

(c) If $V = BD$, find $B(t)$. Use the result from part (b).

$$V = \frac{D}{R}\frac{dR}{dt} = \frac{1}{R}\frac{dR}{dt}D = BD \rightarrow B = \frac{1}{R}\frac{dR}{dt}.$$

(d) We want B. R is increasing at a constant rate of $1.00~\mu$m/s, so $R = R_0 + v_Rt$. At 4 years $R = 500.0~\text{m} + (1.00~\mu\text{m/s})(4)(3.156 \times 10^7~\text{s}) = 626~\text{m}$.

$$B = \frac{1}{R}\frac{dR}{dt} = \left(\frac{1}{626~\text{m}}\right)(1.00~\mu\text{m/s}) = 1.60 \times 10^{-9}~\text{s}^{-1}.$$

(e) We want D. $D = R\theta = (626~\text{m})(\pi/3~\text{rad}) = 656~\text{m}$.
(f) We want the separation speed V. Use the result from part (b).

$$V = \frac{dD}{dt} = \frac{D}{R}\frac{dR}{dt} = \left(\frac{656~\text{m}}{626~\text{m}}\right)(1.00~\mu\text{m/s}) = 1.05~\mu\text{m/s}.$$

(g) We want the time to reach Xibalba. In this universe, $c = 6.35~\mu$m/s and $v_R = 1.00~\mu$m/s. The waves travel along the circular arc, but this arc is increasing in length due to the expansion of space. For an infinitesimal time interval dt, the wave travels through an arc distance cdt and an angle $d\theta = (cdt)/R$. Using $R = R_0 + v_Rt$, we have

$$d\theta = \frac{cdt}{R_0 + v_Rt}$$

Integrating will give a relationship between θ and t.

$$\theta(t) = \int_0^t \frac{cdt'}{R_0 + v_Rt'} = \frac{c}{v_R}\left[\ln(R_0 + v_Rt) - \ln R_0\right] = \frac{c}{v_R}\ln\left(1 + \frac{v_Rt}{R_0}\right).$$

Now solve for t when $\theta = \pi/3$.

$$t = \frac{R_0}{v_R}\left(e^{\pi v_R/3c} - 1\right).$$

From part (d) we know that when the ripple waves are sent, $R = 626~\text{m}$, so $R_0 = 626~\text{m}$. Putting in the numbers gives

$$t = \left(\frac{626 \text{ m}}{1.00 \ \mu\text{m/s}} \right) \left[e^{\pi(1.00 \ \mu\text{m/s})[3(6.35 \ \mu\text{m/s})]} - 1 \right] = 1.122 \times 10^8 \text{ s} = 3.56 \text{ y}.$$

(h) We want the wavelength that the Xibalbans observe. Use Eq. (44.16) with $v = 1.05 \ \mu\text{m/s}$ at the instant the ripple waves are sent, from part (f).

$$\lambda_0 = \lambda_s \sqrt{\frac{c+v}{c-v}} = (1.00 \text{ nm}) \sqrt{\frac{(6.35+1.05) \ \mu\text{m/s}}{(6.35+1.05) \ \mu\text{m/s}}} = 1.18 \text{ nm}.$$

EVALUATE: The received wavelength is longer than the emitted wavelength. This result is reasonable because this "universe" is expanding.